# Richard Hammond, PhD

# THE UNKNOWN
# UNIVERSE

THE ORIGIN *of the* UNIVERSE,
QUANTUM GRAVITY, WORMHOLES,
*and* OTHER THINGS
SCIENCE STILL CAN'T EXPLAIN

NEW PAGE BOOKS
A division of The Career Press, Inc.
Franklin Lakes, NJ

Copyright © 2008 by Richard Hammond

THE UNKNOWN UNIVERSE
EDITED BY KARA REYNOLDS
TYPESET BY EILEEN DOW MUNSON
Cover design by The DesignWorks Group
Printed in the U.S.A. by Book-mart Press

To order this title, please call toll-free 1-800-CAREER-1 (NJ and Canada: 201-848-0310) to order using VISA or MasterCard, or for further information on books from Career Press.

The Career Press, Inc., 3 Tice Road, PO Box 687,
Franklin Lakes, NJ 07417
**www.careerpress.com**
**www.newpagebooks.com**

**Library of Congress Cataloging-in-Publication Data**

Hammond, Richard, 1950–
   The unknown universe : the origin of the universe, quantum gravity, wormholes, and other things science still can't explain / by Richard Hammond.
       p. cm.
   Includes bibliographical references and index.
   ISBN 978-1-60163-003-2
      1. Cosmology—Miscellanea. 2. Physics—Miscellanea. 3. Outer space—Exploration. I. Title.

QB981.H28 2008
523.1--dc22
                                                                                    2007047671

— ✳✳✳ —

# Acknowledgments

I would like to thank Nancy, Katherine, Jennifer, and Matthew, without whose infinite patience this book could not have been written.

Charles Dickens has always been one of my favorite writers, but when I saw the other 27 titles of *David Copperfield*, I began to realize how hard he worked. Because there is no handwritten manuscript for my opus, I thought I would include a few other titles for historical purposes.

*Unknown Universe: Shadows of Shadows*

*Unknown Universe*

*What We Do Not Know*

*Today's Dark Age*

*Nature's Enigmas*

*The Puzzling World*

*Ten Problems That Shake the World*

*Nature's Best Kept Secrets*

*Something Is Rotten in Denmark*

*The Beginning of Science*

*The Mysterious Universe*

*Incognitus Cosmographicum*

*We Next Come to the Question of the Universe, to Which We Do Not Have a Satisfactory Answer*

— ✳✳✳ —

# Contents

– *Chapter One* –

# Cosmic Acceleration

*Darkness there, and nothing more.*

Poe's famous midnight thinker stared into the darkness, seeing nothing, nothing except what his imagination allowed. Now we stare out into the darkness, using every machination of modern technology, but are still limited by our imaginations. The telescope expands our view of the universe to an inconceivable size, and allows us to see into the cold blackness that stretches back to the beginning of time. Today, as we peer into the dawn of the universe, we are bringing back questions from the past like mysterious fossils etched in bedrock. Our most cherished notions about the nature of matter are challenged, and Einstein's famous theory of general relativity, the cornerstone of our understanding of gravity, is placed under the microscope.

These recent observations are not the first to rock our view of the cosmos, but, unlike the previous explosive revelations, these remain unsolved. The ultimate solution may cause us to abandon our most esteemed model of the universe, or may prove the existence of a new kind of matter never seen before.

To appreciate the full effect, let's step back and soak up a little history. It is fun to ask the question, "Where on Earth was helium discovered?" It is fun because the answer is nowhere—nowhere on Earth, that is. Helium was discovered on the sun, even though we never landed there and never will. This is because elements are like hapless criminals, leaving their fingerprints on everything they touch. To understand the details we must take a detour through the fascinating world of spectroscopy.

Start with hydrogen, the simplest atom. If you fill a glass tube with hydrogen and force an electric current through it, the gas will radiate, producing a dull glow. This is precisely what happens in the garish neon signs that emit the familiar sanguine radiance hawking everything from Jell-O to Mr. Peanut.

**Question:**

*Suppose we take this light, from hydrogen, and pass it through a prism.*[1] *What happens?* (I will give you some answers, three of which are wrong, just to get the wheels rolling.)

❑ 1. Nothing.

❑ 2. It is refracted (bent), but still looks bluish.

❑ 3. It explodes into the colors of a rainbow (the same thing that would happen to sunlight).

❑ 4. Four distinct colors emerge, each one refracted by a different amount.

This experiment is repeated countless times every year in every college physics labs, but as the 19th century entered its retirement years, this conundrum had physicists questioning their most cherished notions. It was as shocking as waking up one morning to a green sky. When explanations were ultimately found, it was so hard on some physicists that they refused to accept the new ideas. In fact, this era gave birth to the famous quote, "Physics progresses gravestone by gravestone."[2]

If you gave this question to the physicists back then, before they did the experiment, they would circle number 3 as their prediction. I hear them pontificating, as they stroke their beards, "Sunlight breaks into the rainbow, and so does light from a hot iron poker, or light from a brilliant lightbulb filament, and so would light from any other substance. Get with the program."

Maybe the last sentence is a stretch, but the correct answer knocked their whiskers off. To put it in perspective, recall the lesson of Copernicus and Ptolemy: Old Claudius Ptolemaeus had us believing that everything revolved around the Earth, and no one ever shook things up us as much as Nicolaus Copernicus, telling us we are spinning wildly on an axis. The Church tried the usual—burning people at the stake (Giordano Bruno was burned at the stake in Italy in 1600)— but even its fiery dogma of ignorance was broken down by truth and reality. Copernicus's upheaval was nothing compared to the hydrogen issue. Luckily the Church was unable to burn anybody this time, so the revolution went a little more quickly, stymied slightly by old physicists clinging to their lost reality. (I sometimes worry I am clinging to a lost reality.)

In the actual experiment, the light is passed through a thin slit in order to collimate the beam. For this reason the light that is observed appears as thin lines—exactly four thin lines. If you would like to see the real thing, stop off at the physics department of any university; they will be glad to show you. Just ask to see the spectrum of hydrogen from a gas discharge tube. Or, check out *http://cfa- www.harvard.edu/seuforum/ galSpeed/*.

*Figure 1.1. The four visible emission lines of hydrogen. To the right is crimson, and the others are turquoise, blue, and violet. Each line is a different wavelength of light.*

So why did these thin little lines reduce the Copernican Revolution to just another day at the beach? These lines, along with a few other experiments, gave birth to quantum mechanics, a view of nature that boggles the mind. While allowing us to understand our world, at the same time it severely limits our knowledge of the universe.

The essential point for now is that, according to quantum mechanics, energy is quantized. The hydrogen atom can absorb and emit light, but only by undergoing transitions through a discrete number of energy levels. It is like sitting in an airplane. You can move up or back, but you must take a seat, so your net displacement is quantized. In fact, I once interviewed a hydrogen atom who was kind enough to explain all this.[3] This quantization means that light can only be emitted at certain, discrete wavelengths, which you saw in Figure 1.1. These wavelengths are different for each atom. For example, the spectrum of helium consists of more than four lines and looks nothing like Figure 1.1.

We are making good progress. Now you can understand how helium was discovered on the sun: The spectrum was observed. But that is still not my main point. My main point is this: Elements can be discovered by observing the spectra of their light, and the spectra are unique.

While physicists have continually brought us to ever smaller scales, down to protons and neutrons, and now even smaller, to the quarks that are the building blocks of these teeny particles, astronomers have been making our universe larger and larger. The first modern explosion came with the discovery of an island universe, which I saw one night in Massachusetts, riding back from a lecture given by John Wheeler, one of Einstein's most famous students. I was studying generalizations of Einstein's theory and was very grateful for the encouragement Wheeler gave me. He had spoken at Mt. Holyoke College, and we were driving back to RPI (Rensselaer Polytechnic Institute, in Troy, New York), where I was a graduate student. We stopped somewhere on a peak in the Berkshires (there are no restrooms nearby), and Frank Ferrandino, an astrophysics student, was looking up as we were looking down. He was pointing out the constellations, and showed us how to find Andromeda, which we saw with our naked eyes. You have to know just where to look, and it is faint, but you can tell it is not quite a point as are the other stars. Even with a telescope it is nebulous, a fact good enough to justify one of the early names, *nebula*. Then came the Great Debate. Two astronomers duked it out: In one corner was Harlow Shapley, saying our Milky Way was the whole enchilada, while Heber Curtis was saying it was time to think outside the box. I am taking a little poetic license here; the actual debate, in 1920, took place at the Smithsonian Museum of Natural History, and the idea of another galaxy was tough to swallow back then. Copernicus was hard enough, and now this? Years gave Curtis the victory (see the technical note at the end of the chapter about Cepheid variables), and nowadays we know there are many billions of galaxies, and galaxies such as our beloved Milky Way are more common than candy bars on Halloween. Another name for these blurry spots, as others came to found, was *island universe*—my favorite—but this moniker was also doomed.

This little outing in the Berkshires was quite a moment for me, seeing another galaxy for the first time, but it was broken by an angry motorist, honking his horn to assuage his annoyance. As he passed, the pitch of his horn lowered— an effect heard all the time—but being physicists (or, back then, physicist wannabes), we plunged into a discussion of the Doppler effect, which includes the mathematical formulas that describe the lowering of pitch, or increase in wavelength, as the source of the sound moves away from us.

By the way, and back to galaxies, this brings up another question.

 **Question:**
*How do you find out what these mysterious clouds of light are made of?*

- ❏ 1. Send out a probe, collect some stuff, and bring it back to Earth to analyze.
- ❏ 2. Ask an astrologer.
- ❏ 3. Send out a manned probe, after we settle on Mars.
- ❏ 4. Examine the spectra.

Number 1 would take hundreds of millions of years, if we could do it at all. I will not even address number 2, and number 3 is another dig. Anyway, I know you got the right answer, and astronomers wasted no time doing just that: examining the spectra. They did it for every galaxy they found, and just as one blazing enigma yielded to explanation, more provocative ones ignited in their place.

*Vesto Melvin Slipher* is a name you do not hear every day. I grabbed his mug from the Internet because it is also a face you do not see every day. He was born in Indiana, studied at the University of Indiana, and became famous during the second decade of the 20th century. Well, not that famous, but in my book he should be. He was studying the spectra of light from distant galaxies and found something curious. The spectra looked like known elements, but the lines were a little off. Look at the hydrogen spectrum again. The red line corresponds to a wavelength of precisely 656 nanometers (a billion nanometers is a meter). Slipher saw a wavelength just slightly bigger than 656 nm.

*Figure 1.2. Vesto Melvin Slipher.*

The other lines were shifted in the same manner, each to a longer wavelength. Figure 1.3 shows another example of a spectrum that is shifted. Note that the lines are dark, which means we are looking at an absorption spectrum (see the technical notes at end of the chapter). The lines on the right are moved up toward the red, so we call this redshifted.

Something as extraordinary as this calls for a question, so here it is.

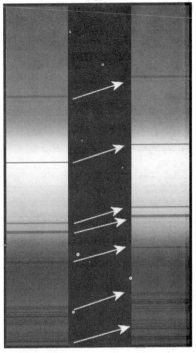

*Figure 1.3. The (absorption) lines are shifted to the red due to the recessional velocity.*

**Question:**

*What causes the spectrum to be redshifted?*

☐ 1. Dirt on the telescope lens.
☐ 2. A mischievous God, like Zeus throwing lightning bolts.
☐ 3. The ozone.
☐ 4. The galaxies are moving away from us.
☐ 5. Light gets tired, traveling zillions of miles and all.

When I taught astronomy, I usually gave easy tests, but every once in a while a few students were determined to undermine my facile approach. Refusing to come to class, they found trouble in the simplest questions, but often relied on "ozone" as the panacea to all issues. I do not know why this is, but that is why I included it—and yes, it is the wrong answer. Number 2 could be argued, but let's try science instead. This question shows how devious I can be—I already gave you the answer! Remember the angry motorist in the Berkshires? The pitch is lowered, or the wavelength of sound is increased, when he is moving away from us. It is called *the Doppler effect*, and it holds for light as well as sound. Once again, number 4 is the correct answer.

Because nothing exists in isolation except isolation itself, I must bring in another thread of the fabric of revolution: Mercury. I am not speaking about the speedy messenger long before FedEx; I am referring to the speedy planet, zooming around the sun once every 88 days. Too small to have an atmosphere, its ravaged surface exposes the callous ferocity of our solar system—but I wander. I want to mention its orbit, which is elliptical. The point of its orbit closest to the sun is called the *perihelion*. One of the greatest triumphs of the human genius, Isaac Newton, discovered the universal law of gravitation and the equation of motion, and proved that Mercury (as all planets) speeds around the sun in an elliptical orbit (nowadays, physics majors discover this in their sophomore or junior years). The theory also predicts that the perihelion is fixed in space and time.

All this amazing theory, corroborated by observations as plentiful as dandelions in my backyard, made many people very happy. But happiness, like a rainbow, never lasts, and soon Mercury proved to be a pesky problem. If I were writing this book in 1910, I would definitely include pesky Mercury as a chapter, for the perihelion of Mercury moves. It precesses around the sun 1,043 seconds of arc every century.

It was not too hard to realize that the other planets whirling around the sun produce a modest tug on Mercury. In fact, calculations show that this effect will produce a precession of the perihelion of 1,000 seconds of an arc per century—1,000 seconds, but no more.

You see the quandary of the old-timers—how on earth to account for those 43 seconds—but this was not the first time astronomers were left in the dark. Years earlier Uranus was known to be an outlaw, breaking the gravitational law of Newton. Times like these are quite fertile for theoreticians like me. We unleash our imaginations, creating new theories and models from thin air, hoping we shed some light on nature's secrets. It was the same back then, and Newton's law of gravitation was quickly put on trial. George Airy, a British astronomer and mathematician with more medals than a four-star general, theorized that Newton's inverse square law did not hold at large distances. After all, just because a theory works on one scale does not mean it applies to all. In fact, when applied to the tiny atomic world, Newton's laws crash and burn like the Tunguska comet (see the technical notes). But there was another explanation for the errant behavior of Uranus: Perhaps there is something else out there yanking the great mass off course. Taking this idea and running with it, Urbain Jean Joseph Le Verrier (born in 1811 in Saint-Lô, France) and John Cough Adams (born in 1819 in Lidcott, England) actually used Newton's laws to determine where such a planet could be.

Armed with hard calculations, William Herschel and sister Caroline swung their telescope where the theoreticians predicted, and, before another perplexing mystery could pop up, it was found: The discovery of Georgium Sidus was made. What, you never heard of it?

*Figure 1.4. Caroline Herschel, sister of William. (Courtesy of NASA.)*

That's because no one could stand Herschel's putative name (except maybe King George), so Neptune took over, following the rule that no planet is allowed two words. The discovery is a fascinating story,[4] but I must get on with my own little story.

Back to Mercury and its errant orbit. What could be a more natural explanation than the existence of another planet? It would be so close to the sun that it would be very hard to see, which explains why we had not. Our friend Le Verrier had such good luck in predicting Neptune, it prompted him to work diligently on the Mercury issue, pondering those 43 seconds of precession. Le Verrier was a terrific mathematician, and he was able to calculate that a planet 13 million miles from the sun (the earth is 93 million) that streaks around once every 20 days would fit the requirements. That same year, 1859, a French country doctor, Lescarbault, who had a modest telescope, saw the planet. By now it was clear that it needed a name. Le Verrier knew the fate of Georgium Sidus, and stuck to the rules this time: He came up with *Vulcan*, and by 1860 awards were flying faster than the new planet, and soon other sightings were published.

You did not know this? Let me pose a question.

**Question:**

*Why is Vulcan, the planet closest to the sun, unknown today?*

☐ 1. It is too small to be considered a bona fide planet.

☐ 2. Le Verrerier needed cash, and created a hoax.

☐ 3. It is classified Top Secret by the Air Force, who know all about it, as well as the little Vulcans that landed in Roswell.

☐ 4. Its discoverers were seeing things.

The definitive answer did not come along until the next century,[5] so let us go to the first decades of the 20th century, when two more famous names arise: Einstein and Hubble. Everyone knows about Einstein, and most people know about Hubble too, because the orbiting telescope is named in his honor. These guys were very active during this period, and Einstein came up with his general theory of relativity in 1915. This theory is a theory of gravity that turns out to be better than Newton's, and more accurate too. In fact, it predicts that the perihelion of planetary orbits *should* precess, and it even predicts the *amount* of precession. Would you like to guess what it predicts for Mercury? Yup, 43 seconds of an arc per century, right on the nose. And that was the end of Vulcan, and no one has seen it since.

Einstein's theory, GR, can not only be applied to planets, but it can also be used to model the entire universe in which we live. This is so cool I cannot understand why everybody does not become a physicist, but experience shows me they do not, so I will go on. During the mid-1920s it was shown that GR can describe a universe that is expanding or contracting (but not static). At the same time, Edwin Hubble was thinking about Vesto's red shifts. Actually, Hubble was doing a lot more than thinking. Working at Mount Wilson Observatory, he made measurements, collected data, and came across the most startling discovery of the century (actually, there are a few "most startling discoveries," but I take poetic license). Here is what he said: "The results establish a roughly linear relation between velocities and distances among nebulae for which velocities have been previously published, and the relation appears to dominate the distribution velocities."[6]

I am no Hubble, but here is how I would say it: "Cosmic secrets have been unearthed, shattering our view of the universe. Observations show that the cosmos is exploding apart, as distant galaxies race away faster than a soccer mom in a van. Even more astounding: the further the galaxy, the faster it goes."

You get the idea. When I taught astronomy the students always liked this, so I'll give you the famous formula Hubble made up:

$$v = H \times d,$$

which states that the velocity of a galaxy is equal to a constant $H$ times the distance $d$ (the distance of the galaxy from Earth). In honor of Hubble we call $H$ Hubble's constant.

Now, finally, the title of this chapter begins to make sense. I am talking about the fact that our universe is expanding. Hubble's data was so convincing, and fit so well with mathematical solutions to Einstein's equation, that it did not take long for this to be accepted as gospel. The universe is expanding because Einstein's equations say it must (Einstein's equation also says it may contract, but Hubble showed us it is expanding). By the way, this only holds for distant galaxies; galaxies that are nearby, such as Andromeda, actually move *toward* us due to the gravitational attraction.

As years went by, more and more galaxies succumbed to Hubble's law. The field of cosmology might have gotten as dull as my garden shears, but eventually outlaws emerged: Distant galaxies, lurking on the edge of our visible universe, are caught loping away from us by astronomers. Like criminals beyond the reach of justice, they are not going as quickly as expected.

I am sure you would not want me to gloss over the related controversy that arose some years ago, so I will not. Until Edwin Hubble, everyone had a firm belief that the universe was a fixed, static thing, allowing motions within, such as planets buzzing around stars, or stars zooming around other stars. This is the household view, in which you and the pooch can roam around, but the structure is firm.

Even Einstein had trouble letting go. He had the temerity to apply his equations to the cosmos, providing a mathematical model of the universe, but the equations did not work! I vividly remember going through this exercise as a graduate student: Where one equation gives the density of matter in the universe, another says the density is zero, which is clearly wrong because we see matter all over. Bummer. In order to obtain a sensible solution, Einstein changed his equations by introducing a cosmological constant. Solutions were found and he was happy. Briefly.

Why briefly? This was before Hubble's explosive revelation, and Einstein assumed the universe was static. What Einstein missed was this: His equations were telling him the universe was not static. He did not hear their plaintive wail, but others did. By assuming the universe can expand (or contract), the equations of general relativity give a beautiful solution without the cosmological constant. These solutions, with Hubble's observations, put the kibosh on Einstein's static universe and his cosmological constant. What did Einstein say about all this? He said the cosmological constant was the biggest blunder of his life. Perhaps he gave up on his pet term a little too easily; you can decide later.

If the universe is expanding, then what was it like before? A million years ago it would have been smaller than it is now, and two million years ago even smaller, and so on, until, at one time, it was no bigger than the period at the end of this sentence. And before that even smaller.

This was too hard too take, for some. Astronomer Fred Hoyle, and others, promulgated the steady state theory, which says that on the large—very large—scale, the universe seems to be homogeneous; the same here as there. Suppose we take this symmetry further, and assume that the universe is the same now as it always was, on average. How can this be, if the universe is expanding? The answer is that matter is being created continuously. Because the universe is so big, you don't have to worry about going into your basement one morning and seeing a new pile of matter. In order to account for the universe as we now see it, you would have to wait centuries and centuries just to see a new electron, which you probably would not notice anyway.

And so it came to pass that there were two camps: the expanding-universe groupies and the steady-state-cosmology fans. How do good scientists convince everybody that they are right? Cast the other view in the pejorative. And so it came to pass that the steady-state advocates made fun of the other side, smirking that the universe began from a big bang. And so it came to pass that this tactic backfired, and subsequent measurements have proven that we live in an expanding universe. And so it came to pass that we had the big bang, and the name was good.

Cosmology was settled, and that is that. Or, that *was* that. As theoreticians such as myself never stop theorizing, observers never stop observing. In 1998 a paper was published that rocked the world. It is the subject of this chapter, and, according to some, represents today's biggest problem. What in the world happened? I would like to quote, almost verbatim, from the 1998 paper. Let me explain first that "SN Ia" are supernovas, stars that explode when gravitational self-attraction makes them collapse. These are the brightest things in the universe, which is why we can see them even though they are incredibly far away. "Different light curve fitting methods, SN Ia subsamples, and prior constraints unanimously favor eternally expanding models with positive cosmological constant...and a current acceleration of the expansion."[7]

So, the cosmological constant is not dead after all?

Let me explain all this. First, remember that when you look into the heavens you are looking into the past. The light from the sun you see right now left eight minutes ago. If you look at nearby stars, the light you see left years ago, and if you look at Andromeda, the light left hundreds of millions of years ago. In effect, the Hubble telescope is able to see billions of years into the past.

What the astronomers really observed is this: Distant supernovas are redshifted. The amount of redshift is proportional to the distance (Hubble's law). The recent observations I quoted show that very distant objects are not redshifted as much as we expected. In other words, billions of years ago the rate of expansion was less than it is today. In still other words, the rate of the expansion of the universe is bigger today than it was in the past, which means that the expansion of the universe is accelerating.

This is so profound it is worth repeating: The expansion of the universe is accelerating. Everybody always thought the rate of expansion should be decreasing. Why? I put it to you.

**Question:**

*Why did everybody expect the rate of expansion to be decreasing?*

☐ 1. The universe, like geezers do, gets tired and slows down.

☐ 2. The intergalactic ozone acts like a sort of cosmic muck, gumming up the works.

☐ 3. It is due to gravity, which is an attractive force.

☐ 4. Nothing goes on forever.

Last week I took a jog. As I was gasping for air, a geezer zipped past me, so 1 is not the answer. Number 2 is my ozone trap answer. Later I will try and think of a question for which *ozone* is the answer, but intergalactic ozone is as sparse as the hair on my head. I am not sure what 4 means, which leaves 3 as the right answer. (By the way, I lifted some of these questions from my astronomy course, so you can keep track and give yourself a grade at the end.)

Suppose you are standing on the moon and throw a rock up in the vacuum. What happens? It slows down, and eventually falls back to the surface. If you could toss it fast enough, it would still slow down on its way up, but it may never return. It slows down because the moon's gravity pulls it back. (We went to the moon to do this experiment so we would not have to worry about the atmosphere.) If the rock started to speed up on its way up, you would have to scratch your helmet, thinking it is a magic rock.

Same thing for the universe, which is made of billions of galaxies. They are flying apart because that is how the big bang started things, but they also tug on each other, so they should be slowing down. Instead, it is like the magic rock we threw, speeding up on its upward trajectory. Something, it seems, is pushing everything away, but the only forms of matter we know about do just the opposite. Let me put it to you again.

**Question:**

*What causes the expansion rate of the universe to be increasing?*

❑   1.   It is not really increasing; the observations are wrong.

❑   2.   It is the cosmological constant: Einstein's big blunder was no blunder at all.

❑   3.   The universe is filled with a mysterious form of matter or energy that we have never seen, and this mystery goo is pushing the universe apart faster and faster.

❑   4.   The general theory of relativity is wrong.

❑   5.   Light gets tired.

❑   6.   The redshifts that lead to this conclusion are caused by something else.

❑   7.   None of the above.

Answer: It could be any of the above.

This is today's zany state of affairs, and no one understands why! This statement is bold, so let me put it this way:

There is no solution that has been generally accepted as the correct answer. Let's go back and reread the statement from Riess and 19 others: *"eternally expanding models with positive cosmological constant."* Now you see why I had to tell the story of the cosmological constant. Although Einstein thought he goofed, one decade's blunder can be another decade's salvation.

Answer number 1 is possible, although Riess and the 19 others would probably not agree. These measurements are very difficult, and there are many intervening interactions that may be involved. I have trouble measuring for a rug, so when the distance must be determined accurately, and an object is billions of miles away, you can imagine how difficult that is. Other observations since 1998 corroborate the result, and, as the evidence continues to pile higher, we might someday be able to rule out number 1. But not quite yet.

If you reread the quotation from Riess (and 19 others) you will see mention of the cosmological constant. Unlike matter, which, through its mass and energy, tends to pull things together, the cosmological constant can act in the opposite way, pushing everything apart. As Einstein knew, it could be a repulsive term, but now it has a much more important job. Unfortunately, this is more of a kluge than a real answer. Why is there such a term? Where does it come from? Why is it so small? (Careful measurements of the orbits of the planets puts an upper bound on the cosmological term, and it is teeny-weeny, if not zero.) Is there any other way to measure it? Is it really constant?

All I can tell you is that this cosmological term may be *part* of the solution, but it is by no means accepted as *the* answer. Answer 3 has gathered enormous attention. This mysterious stuff is called dark energy, and was the original title of this chapter. Dark energy may be natural from a quantum mechanical point of view, in which even a vacuum has energy. Every month dozens of articles appear about dark energy, sometimes called *quintessence*, which can account for a force that pulls our universe apart. This is probably the most accepted view, but it is no real solution, because it creates so many questions. What is dark energy? Where does it come from? Can we measure it here on Earth? How does it interact with matter and light? And so on.

Answer number 4 may go along with number 3. There are many generalizations of Einstein's theory, and some of these contain new fields that may aid the universe in its expansion. I just published a paper that shows how this happens, but basically, the effect is more pronounced in the early universe than the late universe. I have another idea I am working on that seems more promising, but then, theorists always have another idea.

Answer number 5 is also a possibility. Now and then physicists get cute and adopt whimsical titles, like "tired light." The universe is not as empty as it seems: It is filled with electromagnetic energy, neutrinos, and a quite an assortment of other items. The light we see from these distant objects travels through billions of miles of this cosmic soup, and must interact with it to some degree. If the distant light gives up energy through these interactions, its wavelength will change, increasing toward the red. As far as most astronomers and physicists can predict, this is not the solution. If these interactions did occur, then the light would be thrown off course, and distant objects would be fuzzy. They are not.

Answer number 6 goes along with number 5. In fact, many of these may go hand-in-hand. These supernovas are the most distant objects we have ever seen. This means we are measuring things at distances never before tested. It is quite conceivable there are a bunch of effects that on the small scale of galactic sizes do not manifest themselves, but become important as light travels across the vast expanse of our universe.

So, there you are. I am leaving you with this mystery because I have no other choice. Maybe next year someone will realize the distance to these far-off objects is wrong, and that they are really closer than we think. A minor correction to the distance tables and all is well. On the other hand, perhaps our most basic tenets of physics have to be changed, and we are facing a revolution no smaller than the one Copernicus fired up.

What do you think?

## Technical Notes

### Line Spectra

When I explained the line spectrum of hydrogen I was addressing emission spectra, but when we look at the sun we actually see an *absorption spectrum*. The light is generated deep within the torrid regions of the sun, where it is a thousand times hotter than the sizzling surface. As this light leaves the sun, helium and other elements absorb some of this light, but it can only absorb light at the same wavelengths at which it emits light. Thus, we see dark lines where light is absorbed, in the place of bright lines.

### Cepheid Variables

Sometimes we get lucky, and with Cepheid variables we hit the jackpot. These are stars whose brightness varies periodically. The really fascinating phenomena is that the period is proportional to the brightness. Therefore, if you measure the stars' periods, you know how bright they are. So when we measure how much light we receive, we can figure out how far away they are. When Cepheids were found in Andromeda, it was realized that this was another galaxy, hundreds of millions of miles from Earth.

### Tunguska Comet

This baby smashed into Siberia in 1908. It looks as though Tsar Nicholas II was testing atomic bombs, leaving the devastation of a 15-megaton blast (15 megatons of TNT).

*– Chapter Two –*

# Dark Matter

A movie of our solar system might not be much of an action flick, but if you could speed it up 7,603,200 times, Mercury would motor around the sun in 1 second. The Earth would take just more than 4 seconds, and Jupiter would jog along using 45 seconds to make one revolution. Poor old Pluto, the smallest planet, would take forever. (In 2006, the International Astronomical Union destroyed the entire planet. Resolution 6A demoted Pluto to "dwarf planet," but, as you can see, Pluto is still a planet in my book. It takes 250 years to revolve around the sun, which would be nearly 20 minutes in the movie version.)

If you allow me to delve into a bit more detail about this, I will be able to explain the dark matter issue much more clearly. Let us begin with Urania—the muse of astronomy. Perhaps it is no surprise that, when Tycho Brahe built the best observatory in the world (at that time, which was 1580 or so), he named it Uraniborg. Brahe was Danish, and the last of the great astronomers who did not use a telescope. Year after year, Brahe, with help from sister Sophia, made careful naked-eye measurements of the motion of the planets.

Uraniborg, a palace with gardens and state-of-the-art research equipment, attracted many visiting scientists, including the Johannes Kepler. Although Tycho and Sophia had always guarded their observations like the Danish Crown Jewels, they decided to share it with Kepler, who continued observing there for many years, and also began using a telescope. With this information, Kepler formulated the most important law of celestial mechanics. The more I think about it, this is one of the greatest equations of all time.

Before I go on, I would like to point out the foresight of King Frederick II of Denmark (and Norway), who financed the construction of Uraniborg. Without his funding there would be no Kepler's law. Of course, someone would have made the discovery eventually, but how long would we have continued groping in the dark, mystified by the natural wonders that remained incomprehensible? I ask you this because science is still bounded by observation. It takes huge capital to finance well-equipped laboratories, which explains why funding agencies such as the NSF and NASA are so important. (It also takes a few additional crumbs to finance theorists such as myself, who are important too.)

Back to Kepler, who deduced that the planets moved in elliptical orbits. This has always astounded me. Last week I complimented my wife on the new rug she bought (it was hideous, but all husbands learn this prevarication eventually), but she informs me we have had it six months. As you see, my powers of observation are limited, so you can imagine the awe I have for Kepler. He also deduced the famous relation concerning the period *T*, the amount of time it takes to go around the sun, and *R*, the average distance from the sun. Kepler used his trove of astronomical data to discover that the period squared is proportional the distance cubed. I give a few more details at the end of this chapter, but the main thing I want to extract is this: The farther from the sun, the slower the planet goes. For example, Mercury scurries around at nearly 50 km/s (about 110,000 miles per hour), while you and I on Earth amble along at nearly 30 km/s. Jupiter is loping at 13 km/s, and poor Pluto struggles at about 5 km/s. Kepler also knew how to write book titles, and explained all this in his *Mysterium Cosmographicum*. I have seen translations ranging from *Comic Mystery* to *The Sacred Mystery of the Cosmos*, but you can't beat the original. This explains my original orphic title *Incognitus Cosmographicum*, but in keeping with modern times, I stuck to English.

By the way, there has always been a fascinating side to Kepler that I feel duty-bound to discuss. Everyone knew, during Kepler's time, that there were six planets. Of course, later there were nine, but now we are down to eight, and if the International Astronomical Union starts feeling its oats again, we may get back to six. To Kepler, this was a profound issue: Why precisely six planets? He was able to accept his famous equation with great aplomb, but the number 6 hounded him like the Furies. Then, while teaching, he had his great epiphany, but we must take quick detour to understand it.

A fascinating 2,000-year-old-plus fact is this: There are five regular solids, called Platonic solids. These are three-dimensional shapes made of identical faces. The best-known is the cube, made from six (identical) squares. Another favorite is the tetrahedron, made from four triangles. The octahedron has eight triangular faces, like two pyramids glued base to base. The last two are the icosahedron, with 20 triangular faces, and the dodecahedron, with 12 sides, each of which is a pentagon. There are these five, no more, and this ends the detour.

Kepler envisioned placing one inside another, with each being surrounded by a sphere. A sketch from him is shown on page 39.

The planets take their rightful places in the spheres. Because there are five solids, this scheme produces precisely six spheres—one for each planet. Kepler was also able to use this scheme to understand the relative radii of the orbits by placing the octahedron first, followed by the icosahedron, dodecahedron, tetrahedron, and cube.

Nowadays we view this as quaint lore, which is exactly my point. A question of profound interest back then is now

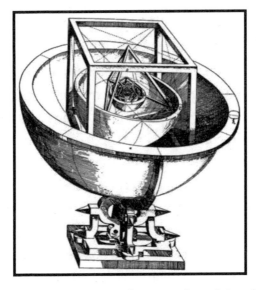

*Figure 2.1. Kepler's geometric framework explains the existence of precisely six planets.*

considered a mere accident of formation. As the solar system collapsed from a great swirling cloud of cold gas and dust, small pockets pulled themselves together from their gravity, forming planets, asteroids, and countless comets. I wonder how many pressing problems of today will be viewed as "quaint notions of the ancients" in the years to come. There are exactly six quarks (called the up, down, strange, charmed, top, and bottom), two of which make the neutron and proton, which form the nucleus of all atoms. Why six? No one knows, but theoreticians are working on it. Will the answer provide some great insight into our unknown universe, or will this question fade like old clothes, handed down from one generation to the next until they are thrown out? Time will tell.

Kepler was not the only person using a telescope back then; so was Galileo. (The Church found Galileo guilty of heresy, but was unable to get him on the stake. Eventually the Vatican recognized his work, but it took nearly four centuries.) When he turned his instrument in 1610 to Saturn he found that the great planet had handles, but his telescope only had a magnification of 20. In 1655, Christian Huygens, who invented the pendulum clock, turned his telescope to this planet, but his magnified the image 50 times. With better instruments come better observations, and with better observations we begin to understand the enigma that is our universe. Huygens proposed that Saturn was surrounded by a solid ring—"a thin, flat ring, nowhere touching, and inclined to the ecliptic."

Huygens published his observation in Systema Saturnium, but the response would make any modern publisher cry. Despite the tepid reviews, speculation and theories concerning the nature of the ring of Saturn abounded for two centuries. James Clerk Maxwell wrote a mathematical essay in 1857 that finally put us on the course to truth. He proved that Saturn could not have a solid ring; instead, it must consist of many independent particles, like a billion tiny moons.

I have left you off the hook long enough, so here is a question.

**?** **Question:**

*Why is it impossible for the ring of Saturn to be a solid piece, like a giant washer?*

☐ 1. Saturn's magnetic field, 8,000 times greater than Earth's, would tear it into tiny pieces.

☐ 2. The solar wind would break it to smithereens.

☐ 3. It would be disintegrated by asteroid impacts.

☐ 4. It's a matter of gravity: It must be in orbit, and the inner parts would rotate faster than the outer parts.

☐ 5. Ozone would build up in pockets and blow it apart.

Saturn's strong magnetic field may wreak havoc on passing spacecraft, but would have little effect on a giant ring. The solar wind is real enough, but by the time it reaches Saturn, it is weaker than a sneeze in a storm. One look at the battle-scarred surface of the moon shows you the devastating power of asteroids, but this takes billions of years. As far as I know, there is no ozone out there, so that leaves, as I know you knew, 4 as the right answer.

I am not just rambling down memory lane here; I am underscoring the fundamental property of orbital dynamics: The further away an orbiting object is, the slower it goes. If an angel swooped by and made a giant solid ring around Saturn, gravitational dynamics would rip it apart, as the inside parts would orbit much faster than the outer ones. Maxwell, in his essay of 1857, used Newton's law of gravity to prove just this.

Now we can move toward the point of this chapter—dark matter—but we must stop off at one of most important years in the twentieth century: 1933. You know that stars like to group together to form galaxies. This is because of gravitational attraction: The stars pull themselves together. By the same line of reasoning, should not galaxies pull together to form clumps? Yes and no; they pull together, but into *clusters*, not *clumps*. How do we know? Telescopic observation shows this. Like people at a party, little groups form everywhere.

Now for some physics. There are two kinds of energy: kinetic, which is due to the motion of an object, and potential. Potential energy is more abstract, and is really stored energy, waiting to be turned into kinetic. Take a spring and squeeze it, and you have potential energy. The Earth has kinetic energy because it is orbiting the sun. The Earth-sun system has potential energy because they are pulling on each other.

The same goes for clusters of galaxies. The galaxies have kinetic energy because they are orbiting around each other, and they have potential energy because they are pulling on each other. There is a theorem that says the total kinetic energy must equal twice the magnitude of the potential energy.

Fred Zwicky, when making measurements of the galaxies in the Coma cluster, found that this formula failed! He concluded that the potential energy was much smaller than it should be.

**? Question:**
*What did Zwicky do?*

☐ 1. He knew right away it was the ozone and did not worry about it.

☐ 2. He realized the theorem was wrong, and made a new one, which today we call Fred's theorem.

☐ 3. He assumed there was invisible matter that accounted for the missing potential energy.

You are probably sick of my bogus ozone answer, but at least you know it is wrong. The next is also wrong, although, as a graduate student, I spent many hours studying the Fredholm Alternative, which is totally irrelevant. This leaves number 3 as the answer, except we call it dark matter.

If there were a lot of matter in these clumps that we cannot see, the potential energy would increase enough so that the theorem holds. By the way, I think astronomers have the hardest job in the world. Experimentalists, for counterexample, can control everything in their labs, from the pressure of the atmosphere to the value of the magnetic field; from the temperature to the density. All astronomers can do is look up, but once again I wander.

A few other similar types of measurements were made throughout the years, but this issue of dark matter did not really steal our attention until the 1970s. I was a graduate student then and did not believe in dark matter. To me, the evidence was thin as spring ice, and I had other ideas. For example, the laws of gravity were never tested on anything as large as the scale of clusters of galaxies before, and perhaps there is some long-range force we do not see on the smaller scale, I mused. My advisor told me to forget about my theory. I should have listened, for within years there was much more compelling evidence.

This evidence is known, within the physics community, as *flat rotation curves*. Remember my spiel about how the outer planets are moving slower than the inner planets, and the parable of Saturn's ring? The same thing goes for galaxies: Near the center of a galaxy the motion of the stars is complicated, but as you move out, almost all of the mass is inside the orbit of a star, and Kepler's law should hold. This means that, eventually, the further the star is from the center, the slower it must go.

Astronomers are good at measuring the speed of stars; they measure the redshift, as we discussed in Chapter 1. They can even measure the speed of hydrogen gas, due to radiation it emits. By measuring the hydrogen emissions, they can make observations way out, even beyond the edge of the visible galaxy.

Astronomers make plots of the velocity of the stars (or gas) versus distance. Here is a little plot I made up for our solar system. It gives the velocity of a planet in terms of its distance from the sun (assuming circular orbits). The distance is in AUs and speed is relative to the Earth's, which is about 30 km/s.

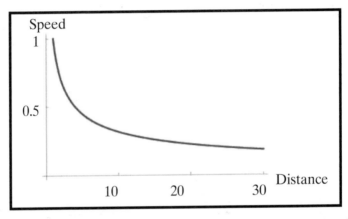

*Figure 2.2. Velocity of the planets (relative to Earth's orbital speed), versus distance (measured in AUs).*

For a galaxy, once you get beyond the busy innards where stars buzz around like a pack of angry gnats, the curve is expected to be the same shape. The key feature is that the velocity gets smaller and smaller as you get further away, which is shown by the downward-sloping curve. It is called *Keplerian*, because it is what Kepler's law predicts. It works great for our solar system, but fails miserably for galaxies. In fact, the rotation curve is flat, indicating that the outer stars or gas are moving much faster than theory predicts.

**?**

**Question:**

*Why does Kepler's law fail for galaxies?*

- [ ]  1.  George Airy was right after all: Gravity gets weaker at great distances.

- [ ]  2.  There is not really enough data to be sure about all this.

- [ ]  3.  There is invisible matter that exerts an additional force on the outer stars, making them go faster.

- [ ]  4.  We have no business applying out prosaic terrestrial laws to something as grand as an entire galaxy.

Number 1 is wrong, and 4 can be ruled out by popular opinion. We are swimming in more data than a river of trout, so 2 is out, which leaves 3—dark matter. Following is a real rotation curve taken from some literature on the subject.[1] Let me help you out with a couple of technical details first: A light-year is the distance light travels in one year, which stretches further than a politician's spin (1.6 billion miles, in more tangible units). A parsec is about 3.1 light-years, and a kiloparsec, or kpc, is a thousand parsecs. Our galaxy, as are many, is about 30 kpc in radius. The next detail I should mention is about names. In the old days galaxies were given great names, such as Andromeda and the Large Magellanic Cloud, but, as telescopes got bigger, the number of galaxies grew faster than Kansas corn, so the New Galaxy Catalogue was created, and each galaxy was reduced to a number. The majestic Andromeda is now reduced to NGC 224.

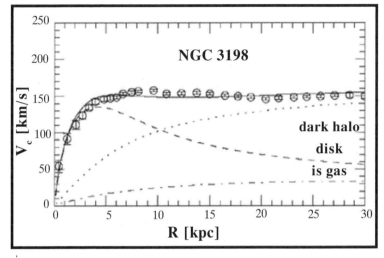

*Figure 2.3. Rotation curve for a real galaxy. The date (circles) stands in stark violation of the theory (downward-sloping dashed line).*

The circles come from measurements, and, as you can see, sketch out a flat line. The dashed curved that slopes down is the Keplerian curve, clearly at odds with observation. Unlike the solar system, the curve rises for the first 5 kpc or so, which is because we are still in the dense region near the galactic center. Once we reach 10 kpc or so, most of the visible mass is contained within that radius, so the curve should start sloping down, as the dashed line shows. It does not. Moreover, rotation curves such as these have been observed for hundreds of galaxies by observatories around the world.

It is generally agreed that these enigmatic curves, the flat rotation curves, are due to the presence of dark matter. The precise definition of dark matter is this: Dark matter acts gravitationally just like other matter, but we cannot see it. You can see from the curve in Figure 2.3 that this is not a small effect; the Keplerian curve falls away from the data faster than an anchor on a broken chain. It is about half the value it should be at 30 kpc, and gets worse after that. This means there must be a boatload of dark matter, not just a few burned-out stars that skulk around unseen. In order to explain these flat rotation curves, it would take up to 10 galactic masses worth of dark matter littering the galaxy! This means that somewhere in the neighborhood of 90 percent of the mass, or more, in a galaxy is dark matter.

This, in turn, means that all of our theories, all of our great physics, are based on observing just a small minority of the matter in the cosmos. How could we possibly have gotten everything right? We do the best we can, but we cannot say we are done. The science community would gladly give all the moon rocks we own to get a few handfuls of dark matter, and we are beginning to speculate on the kinds of experiments that might shed light on this dark subject.

But what *is* dark matter? And do not let me get away with something as wishy-washy as *dark matter acts gravitationally just like other matter, but we cannot see it*. From the 1970s on there has been a great back and forth: Suggestions come and go, some are still with us while others are dashed like a shipwreck on the Carolina coast. I will tell you some of these ideas, and leave you with our best current guess.

One more detail first: The fact that the rotation curves are flat means that the density of dark matter has to be adjusted to give this result, so that the laws of physics still work. It turns out that the density of dark matter must fall off inversely as the distance from the galactic center. I am telling you this, (1) so that you can sleep tonight, and (2) for other reasons that will become clear in the next chapter.

When I was a graduate student and first heard about dark matter, I said, "Big deal. It's probably hydrogen." If that were correct, I would not be writing this chapter. Until dark matter came along, hydrogen was the most abundant substance in the universe, so it seemed logical that it could infiltrate the stars and fill a galaxy. But why doesn't it fall to the center, what maintains its special density, and why doesn't it collapse into stars? The answer is that it must be hot enough to keep itself spread so wide, and if it were that hot we would see it, because it would emit energy. But we do not see it, so hydrogen is ruled out.

This sort of back-and-forth has been going on for 30 years, and is part of the reason it is so much fun to be a physicist, but it is time for another detour, which I will start with a question.

---

**Question:**
*What will happen to the sun in five billion years?*

❑   1.  Nothing. It will live on in perpetuity.
❑   2.  It will turn into ozone.

☐ 3. This question is too hard. Nobody knows.

☐ 4. It will puff out, becoming a red giant so big that it will contain Earth, then it will cool off, by which time it will be a great mass of carbon. As it cools, it will disappear from sight.

Number 4 is the answer. In fact, all stars that are about the mass of our sun will suffer this fate. Some already have. If you look at the constellation Orion, you will see that one of the corner stars (top right) is red. This is Betelgeuse (pronounced BET-el-jooz; many people just say "beetle juice"), which is in the red-giant phase.

The point of that little detour was to explain that many stars eventually follow the footsteps of General Douglas MacArthur—they fade away. Even after fusion stops, stars will glow, but in time, like a piece of coal kicked from the furnace, they will cool off and fade from sight. So, maybe galaxies are filled with burned-out stars? Maybe, but the problem is time. It takes a long time for that many stars to go through the process, and cool down so that we cannot see them. Old, dead stars could be the stuff dreams are made of, but may not be dark matter. Scientists have made ingenious observations that may turn this view around, but so far we cannot find enough evidence of this view.

What's left? Another detour, where we find ourselves knocking back moonshine and bathtub gin. It is the 1920s and we will drink in the story of the little neutron. Back then physicists knew that atoms where made of neutrons and protons—squeezed into the tiny pocket of a nucleus—and electrons, buzzing around much further away. They also knew that a neutron could explode, leaving behind a proton and an electron. We call this *beta decay*, but there was a problem (and yes, if I were writing this book back then it would be Chapter 1).

If you have 1,000 free neutrons (free as opposed to stuck in a nucleus), and you wait 15 minutes, what do have? You have 500 neutrons (technically we describe this by saying the half-life of a free neutron is 15 minutes). Back in the roaring '20s physicists made careful measurements of this decay process (and still are), and were dumbfounded. The mass of a neutron is a little bit bigger than the mass of the proton plus electron. This difference in mass, through Einstein's famous formula $E = mc^2$, shows up as kinetic energy of the electron. Conservation of energy makes this absolutely true, and the electron must zip away at exactly 0.59 times the speed of light. Even a team of California lawyers cannot argue this away. There is no doubt, no room for negotiation—the electron must motor off at exactly this speed to conserve energy. The problem was, the electron was measured to have speeds less than this value! It was measured to crawl away like an inchworm after lunch, or dart off at more than half the speed of light, and anything in between.

**?**

**Question:**

*What was the solution to the case of the errant electrons?*

❏ 1. Some other wraithlike particle no one ever saw carried away the energy.

❏ 2. Energy is not really conserved, not on this tiny scale.

❏ 3. Beta decay is really triggered by an unseen particle, which accounts for the energy difference.

Enrico Fermi was one of the truly great physicists. He was raised in Italy and came to Columbia University in 1939. He had been thinking about this problem and heard Wolfgang Pauli's idea of another neutron. Fermi liked the suggestion that the electron had an accomplice, an unseen particle carrying away energy. But it would have to be neutral like the neutron,[2] and very light (otherwise conservation of momentum would be violated), so Fermi called it *a little neutron*, and worked out his theory. But he was speaking in Italian, so the particle became the *neutrino*. A year ago I was reading the *Wall Street Journal*. I do not always agree with the writers' opinions, but they were talking about neutrinos and called them "wraithlike." I thought that was a great choice of words, and freely use it now: Number 1 is the correct answer, and, though no one had seen it, the neutrino was born. These particles interact very weakly with matter and are hard to detect, which explains why it took 20 years to experimentally detect the neutrino. To demonstrate this, I ask:

**Question:**

*An enormous number (more than $10^{12}$, which is a million times a million) of these particle pass through you every second. What are they?*

❏ 1. Neutrinos.
❏ 2. Ozone ions.
❏ 3. Photons.
❏ 4. Cosmic rays.

Yep, neutrinos, coming from the sun. In fact, they pass right through the Earth itself. I was not kidding when I said they interact weakly with matter.

Okay, the universe has more neutrinos than traffic in New York City. Although they carry energy and momentum, we used to think they were massless (see the technical note about massless particles), but we changed our minds, and some people began thinking they were the perfect candidate for dark matter. Well, not perfect. It is hard to understand how their density in a galaxy can vary as I described. They are so light that they travel very nearly equal to the speed of light, and they would not tend to clump in galaxies, but careful observations show that dark matter *can* clump, So, neutrinos may be part of the dark matter, but there must something more.

Which brings me to WIMPs, or Weakly Interacting Massive Particles. The name says it all. Something like neutrinos, they barely interact with matter but have mass, perhaps a lot of mass. By the way, what is the difference between a hot cup of coffee and a cold beer? If you stick your finger in, all you can tell is that one is hotter, so what is the difference to your finger? The molecules in the hot coffee move, on average, 11 percent faster than those in cold beer. In other words, the hotter something is, the faster the particles move. This prompts physicists to break dark matter into hot and cold: Neutrinos, or any very light particles, are hot dark matter, because they move so fast. Heavy unknown particles that slug along like sap from a Vermont maple, are said to be cold dark matter.

Back to WIMPs. It is now believed that, because WIMPs may not clump, there must be other dark matter, and so came MACHOs—MAssive Compact Halo Objects. These things could be objects that collapsed but were not big enough to form a star, such as Jupiter. They could also be stars that burned out, and are too cool to see, and maybe a few black holes thrown in for good luck, but I will talk about black holes in the chapter on quantum gravity. These are considered cold dark matter, and some models of dark matter contain a mixture of hot and cold matter. The trouble with too much cold dark matter is that we should see some evidence of it. As light works it way through the MACHO-strewn universe, it will suffer small deflections—microlensing—from the gravitational field. Although there have been reports of microlensing, the evidence is still too thin to be definitive.

Often in physics a problem is really a solution in disguise, and this might be the case with the axion. The axion is a particle that is postulated to exist and weigh between $10^{-7}$ and $10^{-11}$ times the mass of the electron, making it very light indeed. I describe the rationale behind why we believe this particle could exist in Chapter 9, but, although we have been searching for decades, it has remained as hidden as Atlantis.

Axion searches have been underway for decades, and, like Sherlock Holmes with his magnifying glass, physicists have been sifting through the results of high-energy accelerator experiments looking for a clue that this particle is real. Likewise, astrophysicists calculate the effect such a particle would have on star formation and other astrophysical phenomena, while cosmologists ponder axion physics of the early universe. These rigorous contemplations are what give us the mass bounds I quoted earlier.

In addition, there are experiments we can do right here in River City. Although the axion will decay into two photons, we would have to wait around for longer than the lifetime of the universe. We can help the process along by placing the axion in an electromagnetic cage tuned to just the right frequency. This experiment is like turning on your microwave and waiting for the axions to decay, and then measuring the photons coming out. How did the axions get in there in the first place? If axions represent dark matter, then they account for 10 times the amount of matter we see, so they should be everywhere. Even though it has never been observed, to many the axion is the favorite horse in the race for dark matter, and I will have a lot more to say about it in Chapter 9.

As I forewarned, as soon as one kind of dark matter is brought on stage, careful analysis usually cuts its run shorter than an off-Broadway play (except, maybe, for the axion). So, some began to speculate that the solution is not dark matter at all. Maybe the force law we use to calculate the orbits is wrong. Until recently, I must admit this is what I really thought. For example, let us look at the lesson of pesky Mercury. Why is Mercury singled out as worst offender of Newton's laws? Because, being closest to the sun, it experiences the strongest force. When laws of physics are pushed beyond the limits for which they are known to work, they often fail. The same thing may be true for great distances. The laws of physics have never been tested at this great scale, and it is theoretically possible that there is a long-range force, which we do not see on the "small" scale of our solar system, that is responsible for the whacky motions of remote galactic material.

Following this notion is MOND, or MOdified Newtonian Dynamics. This approach tinkers with Newton's second law of motion so that it produces the correct rotation curve without introducing dark matter. It is a blatant fudge factor, but sometimes that is how science progresses. Later, theoreticians put MOND on a firm theoretical basis, and it seemed to be a possible explanation of the observations. I was not particularly thrilled with this theory, but I liked the underlying ideas of looking beyond our existing laws of physics.

Then, one day, as I was scribbling away, I heard,

"Oh, Dear, they found dark matter."

"Thank you, Dear, can you bring some in?"

"I'll get it, Dear."

My wife swept into my study with an article from the *News and Observer*. It referred to this headline: "NASA Finds Direct Proof of Dark Matter" (see *www.nasa.gov/vision/universe/starsgalaxies/dark_matter_proven.html*).

The picture accompanying the article showed gravitational lensing. I will tell you more about this later, but for now I will simply say that light bends around massive objects. A cluster of galaxies is a massive object, so light coming from behind it will bend around it. The amount of the bending is proportional to how much mass it skirts. By measuring this angle, astronomers can infer how much mass is there, but because we do not see anything, we call it dark matter.

It is difficult to account for this by tinkering with the laws of gravity (as I often do), so many are convinced there is dark matter instead. Not everyone, though, is convinced. John Moffat is a Canadian physicist who has worked many years on examining the mathematical physical consequences of generalizations of Einstein's theory. In his nonsymmetric theory of gravity he provides a gravitational basis for the flat rotation curves, and even claims to explain the data in the NASA picture.[3]

This state of affairs reminds me a little of Uraniborg. The ancient observatory opened new windows to our universe, allowing us to see the world we live in a little more clearly. The Hubble telescope and related scientific instruments, in my view, are an equally great investment. We are beginning to see more of our universe than ever before, and for the first time we may be embarking on the path of true discovery.

Let me explain this explosive statement. Every atom has a nucleus, and nuclei are made of neutrons and protons—but you knew that. We call these particles *baryons*. The quarks that make the neutrons and protons are also baryons. In fact, any particles that are involved with the strong nuclear force responsible for holding nuclei together, are baryons. Everything we see and touch is made of baryons, so if you are having a mug of coffee, you can say you are drinking liquid baryonic matter. If you are wearing clothes you might say you are encased in conformal baryonic matter, and so on. (Electrons are not baryons; they are called *leptons*, and of course go along with the baryons.)

Why the Brutus routine on baryonic matter? To lay the foundation for Antony, of course. Now that you are convinced everything is baryonic, I will try to expand your views, and perhaps you will accept the existence of non-baryonic matter. It seems that almost every form of baryonic matter gets ruled out as a candidate for dark matter, so people began to consider the idea of non-baryonic dark matter. In fact, many people are quite convinced that dark matter is non-baryonic.

Will this mystery be the impetus for research and discovery? NASA is planning IBEX (Interstellar Boundary EXplorer), a mission to the edge of our solar system, where pockets of dark matter may exist. What amazes me is the cost of this mission: At $134 million, it costs less than a week at war, and brings only positive consequences. At the edge of the solar system, detectors will be on the lookout for the kinds of particles that lurk in the interstellar medium. Maybe they will find nothing more exciting than the prosaic proton, or the enduring electron, or maybe a new form of matter will be discovered. Maybe this will be the first shot of a revolution no smaller than the one Brahe and Kepler fired.

For a moment I must leave you in the dark. We do not know what dark matter is. We do not know from where it comes. We do not know how it is made. We do not how it interacts with matter (except gravitationally). We do not know what it looks like, and so, no, we do not know much. I will describe one class of dark matter, a class that some physicists believe is the natural solution to our dark quandary, but that will take an entire chapter. This will be Chapter 8, on strings, so, all I can say is read on.

## Technical Notes

### Kepler's Famous Law

If we measure the period in years and the distance in astronomical units (an astronomical unit, AU, is defined as the distance between the Earth and the sun, so 1 AU = 93 million miles), this equation takes the form $P^2 = R^3$. For example, $R$ for the Earth is 1 (remember: we are using AUs), and $1^3 = 1$, so $P$ is 1, or one year. For Jupiter, $R = 5.2$ AU, so Kepler's law gives P = 11.9 years. You can check that $(11.9)^2 = (5.2)^3$, to within round-off error.

### Massless Particles

Light comes in little bundles of energy called *photons*. Photons carry energy and momentum but have no mass. This means that they must travel at the speed of light. You cannot stop them. If absorbed, they disappear, and their energy and momentum are transferred to whatever absorbed them. Later I will mention gravitons, which we expect to be massless also. We used to think neutrinos were massless, but now we believe they have a few thousandths of the mass of the electron.

– *Chapter Three* –

# The Cosmic Ray Paradox

You have been very patient, and should be rewarded. To make you buy this, I have a surprise—the kind that has a sneaky unknown clue hidden somewhere in the universe, but you must find it today. Sometimes you should read between the lines, and sometimes you must read between the words. Sometimes, you will even totally skip five words. Be observant, and you will see great things happen.[1]

I am sure you can figure this out, so I will apologize for the blatant plug, but often this is how we see the universe. Clues to nature are strewn about like flowers in the Appalachians: persistent, ubiquitous, usually quiet, and always beautiful. But we have to stop and smell them. Cosmic rays were like this, zipping through every lab in world, just waiting for someone to decipher their significance. Let us step back in time to examine some of these clues nature has been so generous to provide.

The largest and most luxurious ship ever built, deemed unsinkable, left England on its maiden voyage to New York. In Savannah, Georgia, Juliette Gordon Low organized 18 girls in order to bring them out of their cloistered lives, forming the Girl Scouts of America. As the greatest athletes from around the globe met in Stockholm, Sweden, electric timers were introduced in order to time the blazing speed of the runners of the Olympics. It was 1912, and in that same year Victor Hess sent a balloon 5,000 meters into the atmosphere, forever changing our view of the cosmos and winning him the Nobel Prize 25 years later.

The problem that had been bothering scientists was static charge. If you walk across a carpeted room on a dry winter day and touch the faucet, you risk being zapped by the electrical discharge. If you cannot stand the jolt, you can be more circumspect about your movements—and the scientists of the time were very circumspect. They were not worried about the dreaded sting of a spark on their finger; they agonized over their instruments. Even though they insulated their detectors so that no charge could flow into them, in time they would become charged anyway. How? The physicists looked at cosmic rays the way today's developers see trees: a nuisance they must expunge.

How can a neutral object, sitting untouched and totally insulated, become charged? Puzzling, no? In physics, sometimes it takes one puzzle to solve another puzzle, so let's turn the calendar back just a few more years to 1895. Wilhelm

Roentgen was studying electricity in rarefied tubes. The electric current created mysterious rays that could travel through paper, and even penetrate flesh and blood, but were stopped by metal or bone. These spooky emanations were named *x-rays*. Within a year Henri Becquerel began to wonder if this had anything to do with uranium salts, which would shine in the dark after being exposed to sunlight. In 1896 Becquerel discovered that, even without being exposed to the sun, uranium could fog photographic plates that were sealed and covered.

All these little clues brings us to the importance of being Ernest—Ernest Rutherford. He realized that uranium emitted different kinds of rays: alpha rays, which are positively charged, and beta rays, which are negatively charged. Like cabalistic prophecies emanating from the Oracle at Delphi, bizarre particles continuously spew forth from uranium. Luckily, Rutherford was able to decipher the fey radiation, and soon Paul Ulrich Villard discovered yet another kind of energy that uranium emitted: a neutral ray. In keeping with Rutherford's alpha-beta system, he named them *gamma rays*. All these names are still with us (Bequerel named the energy from uranium *U-rays*, but this name had a very short half-life), but now we know that alpha particles are helium nuclei (two protons and two neutrons), beta rays are electrons, and gamma rays are photons, or light energy. (X-rays are photons too, but of a lower energy.)

It seemed possible that the problem keeping Hess up all night could be explained by these strange new rays. Uranium, and perhaps other radioactive materials, could be distributed throughout the crust of the Earth. Like submarines lurking beneath the surface, they could silently emit their charge from the shroud of anonymity. And so it was reasoned that natural radioactivity in the Earth sent charged particles through the air and into the scientists' delicate instruments, leaving them with a net charge. Even with this putative explanation, scientists doubted there was enough radioactive material to account for the charging, and the controversy among physicists became as charged as their instruments.

Hess suspected that the puzzling rays came from space. To test this hypothesis, he reasoned that, if he sent up a detector 3 miles into the sky, there would be more radiation, because there would be less air to absorb the energy. In fact, beta rays (electrons) only get a few centimeters before some adoring molecule gloms onto them.

Last year I was in Albuquerque at the University of New Mexico to review a program on high-powered fiber optic lasers funded by the Joint Technology Office. I was at a coffee shop across from the university, and I was trying to convince a cop not to tow my car away (the towaway signs always give away the best spots to park). Just as our "discussion" was warming up, a flotilla of color swooped across the sky, exhibiting one of Albuquerque's favorite sports: hot-air-balloon races. The orb-shaped palettes hissed occasionally as the pilots turned their valves, and the technicolor balloons raced north, ignoring the little controversy below.

In 1912 the balloons were different, but the physics was the same, and up they went. Hess placed his radiation detector (see Technical Notes) in the inflatable device and watched with awe as the radiation waned for the first few hundred feet. But as the balloon ascended into the heavens the face of puzzlement changed to the face of contentment—the radiation grew stronger. By

*Figure 3.1. Victor Hess in his balloon (from* www.fnal.gov/pub/inquiring/timeline/04.html*).*

the way, this emprise occurred long before instruments could send their data back to the lab via radio waves. As the balloon ascended into the rarefied heights 17,000 feet above the Earth's surface, where the oxygen thinned and the temperature plummeted like a convict's expectations, Victor was right there to enjoy it all.

Hess was not working in a vacuum; others had been doing related experiments, some of which used the Eiffel Tower, but this experiment/demonstration is a good demarcation for the discovery of the mysterious rays of unsure origin. The balloon experiment showed that they come from outer space, not the Earth.

But what exactly are these rays, people wondered. Charged particles could be accelerated by electric and magnetic fields, but these particles were extremely energetic, and no one could explain the origin of particles with such brio. Gamma rays, though neutral, could easily cause neutral objects to be charged by ripping electrons away from the atoms, or bursting the nucleus asunder, but no one could understand where such energetic gamma rays came from either. Like a good mystery novel, the ensuing years made the plot thicken. By the time the Great Debate took place between Curtis and Shapley in 1920, two physicists were preparing to square off on this issue—which calls for a slight detour.

The most difficult lab experience I ever had was the Millikan oil drop experiment: In a darkened, tenebrous room, miniscule droplets of oil were sprayed into a small enclosure with an electric field as, through a microscope, I spied their fall. Some droplets may have one unit of charge, some two, and so on. This gives the drops different accelerations and velocities. From this, one can deduce that charge is quantized, and actually compute what that charge is. I say "one can deduce," but that "one" is not me. I could not see a thing, even though we used polymer spheres, which are easier to see than tiny drops of oil. After an hour in the dark I had red eyes, a sore neck, and bad language. This is another reason I became a theorist.

Robert Millikan, who used oil, invented this form of torture in 1909, and was able to compute the charge of the electron better than all previous work, winning the Nobel Prize 14 years later (no pain, no gain). Meanwhile, Arthur Compton took seriously the notions that Planck and Einstein espoused, that light consists of particles—photons. Compton did a calculation that was so simple, so beautiful, I had no choice but to become a physicist. He considered the problem of light interacting with an electron. Before that, everybody used the wave equation of electrodynamics that gave a long and tedious calculation—it also gave the wrong answer.

One of the traits of great physicists is, once they believe in something, they take that idea and push it for all it is worth. Newton applied the laws of gravity to the entire solar system, a bold step back then, and Einstein applied his equations to the entire universe. If light consists of particles, Compton argued, then treat it like particles. He described the collision between a photon and an electron precisely as we would describe the collision between two billiard balls. With this, he correctly described the scattering of light by electrons, and the process and formula are now referred to as *Compton scattering*.

By 1925 these two greats squared off in the ring of scientific debate: Millikan maintained that the particles that came from outer space, which he dubbed cosmic rays, were photons, while Compton subscribed to the notion of charged particles. One of the early telltale properties of this radiation is that it changes with the magnetic field of the Earth. Photons are oblivious to the field, but charged particles feel a sideways force that is proportional to their velocity.

*With better instruments come better observations, and with better observations we begin to understand the enigma that is our universe.* As the ominous sandstorms of the vast Dustbowl grew quiet, and the Great Depression gave way to the Second World War, the evidence began to pile up. Cosmic rays, the subject of this chapter, came to be accepted as charged particles. Anything from a proton to an iron nucleus would do, but probably not photons.

Cosmic rays are more than the stuff of which a chapter is made; they hit us all day long. In fact, the annual dose is about what you get in a mammogram. The difference is that cosmic rays have a vast range of energies, from about the same as a particle in uranium decay to more than a million million times more energetic. Aden and Marjorie Meinel, retired astronomers, claim that ice records from 40,000 years ago show an increase in cosmic radiation, a time when significant evolutionary changes occurred. Others claim that cosmic rays affect climate. These theories are not accepted as scientific fact, but they do make you wonder just what cosmic rays can do. In fact, lately, whenever I make a mistake, I attribute it to a cosmic ray ripping though my neurons, but I am sure the flux of cosmic rays is not high enough to explain every mistake.

Cosmic rays span a vast range of energy, so I need to take another detour to discuss this. Here are a few examples to give you a feeling about the joule, the unit of energy in SI (Standard International) units. If you raise 1 gram of water 1 degree celsius, it takes 4.2 joules of energy. It takes 1 joule of energy to lift 1 kilogram (2.2 pounds) 4 inches off the floor. A 2-kilogram blob traveling at 1 meter per second has a kinetic energy of 1 joule. A kilogram of TNT releases just more than 4 million joules when it explodes. A 100-watt bulb emits 100 joules of energy each second. (See also the Technical Notes at the end of the chapter).

When an electron undergoes a transition in an atom, corresponding to the line spectrum we saw in Chapter 1, the energy emitted is on the order of $10^{-19}$ J. Such a tiny number enfeebles the majesty of atomic transitions, so we use a more convenient unit, the electron volt, or eV, which is equal to $1.6 \times 10^{-19}$ J. Nuclear transitions heave off particles with typical energies of millions of eVs, or MeVs. Great accelerators are built to make particles go very fast, such that their energies are a thousand MeVs, which we call a GeV. The table following is a rough and approximate guide.

| Energy | Process |
|---|---|
| eV | Atomic transitions, visible light |
| 1000 eV = keV | Medical x-rays |
| 1000 keV = MeV | Gamma rays, nuclear transitions, solar cosmic rays |
| 1000 MeV = GeV = $10^9$ eV | Gamma rays, particle accelerators, mass energy of protons, solar cosmic rays |
| 1000 Gev = TeV = $10^{12}$ eV | Cosmic rays, future accelerators? |
| 1000 TeV = PeV = $10^{15}$ eV | Cosmic rays, explainable as coming from a supernova |
| 1000 Pev = EeV = $10^{18}$ eV | Cosmic rays, about 1/10 joule, UHECRs (Ultra-High Energy Cosmic Rays) |
| 1000 EeV = ZeV = $10^{21}$ eV | Cosmic rays, major league fastball, UHECRs |

*Figure 3.2. Energy Table.*

One of the questions we have been wrestling with for a quite a while is, how are these cosmic rays made? And from where do they come? And how do they acquire such huge energies? The best explanation for some of these particles is that they come from a supernova explosion. Supernovas are the most important things in the universe to us—they give us substance. We often think of stars such as our sun spewing out more energy than a roomful of toddlers, but for larger stars we should think of them as automatic alchemy machines. Let me explain.

Not long after the universe began, it was mostly hydrogen, peppered with a little helium. And that's it. So where did all the iron and oxygen come from, not to mention gold and platinum, and everything else? After a billion years or so, great clouds of hydrogen collapsed, from gravitational attraction, into searing orbs. At the centers, temperatures soared, easily reaching 15 million degrees for our sun. At this temperature the nuclei of the hydrogen (protons) collide. It is a raucous party for the particles, and, after spending so many millions of years in the dark gloom of isolation, they really whoop it up. When the dust settles, two hydrogen atoms annihilate, and a helium atom is formed. This is fusion. (I am leaving out some details about how the neutrons come into the picture.) Every time a helium atom is formed, energy is released. It is 5 billion years' worth of this energy that allows me to write this, and you to read it.

I give more details in the Technical Notes at the end of the chapter, but without doubt, this is one of most fascinating aspects of our universe: The helium nucleus consists of two protons plus two neutrons, but the mass of the helium nucleus is *less than* the mass of the two protons plus the two neutrons that made it. Where did the mass go?

As I think about it, this may not seem unnatural to everyone, but physicists live and die by conservation laws. I already mentioned conservation of energy, but in 1905 Einstein derived the most famous formula in all of physics, $E = mc^2$. This formula tells us that mass does not disappear from the universe, but can change into energy, usually as photons or kinetic energy, or both. The difference in mass is converted into energy (see the Technical Notes for details), and is responsible for all of the energy we receive from the sun.

(By the way, now that I stop to think, we all live by conservation laws. For example, if you buy a box of breakfast cereal, you would not be overjoyed to read on the side, instead of weight, the number of flakes. What is better, a box with 3,141 flakes, or a box with 1,414 flakes? By the time you get it home, the number of flakes can double, especially if you drive a tiny car with a trunk the size of a glove box, as I do. Unlike the number of flakes, weight is a conserved quantity. So you see, physicists are no different from consumers of cornflakes.)

After a while (10 billion years for our sun), there is no more hydrogen in the hot core, but fusion continues—this time three helium atoms fuse into a carbon atom, until the helium is gone. For our sun this is curtains, but in heavier stars fusion can continue, crunching its way through neon and selenium all the way to iron. Fusion cannot continue beyond this; to make a heaver atom, such as gold, it requires energy. For nuclei lighter than iron, the attractive nuclear force greatly overpowers the repulsion of the protons, which is why fusion gives off energy, But for heavier nuclei, the repulsive electric force becomes so great that its fusion would absorb energy. Conversely, if the heavy atoms break apart, they give off energy (see the Technical Notes on fission vs. fusion). This is a well-known property for power companies and bomb-makers: fission.

This is why I called stars *alchemy machines*, but what about the heavier atoms, and how on earth do these elements get to Earth? The answer lies in the second part of the story, which describes stellar evolution after it has turned into a giant hunk of iron. At this point, the gravitational forces are so overpowering that the star collapses into a giant neutral nucleus. The electrons are squeezed into the protons, and the collision results in the formation of neutrons and neutrinos. This is a very strange object, yet a common object as we scan the skies. It is called a *neutron star*, but there is a third part of the story.

If the giant nucleus is more than a couple of solar masses, once again the gravitational force is unstoppable, and the star collapses. What is left behind is yet another story, but as the collapse takes place, matter is squeezed much closer together than it can stand, creating the most violent explosion in the universe. If you drop a tennis ball, it is also squeezed together more than it can stand, which explains why it rebounds. With a collapsing star we are talking about nuclear energies, much bigger than the electronic forces of the tennis ball, more than $10^{30}$ times as much mass. During the stellar collapse and explosion there is plenty of energy to go around, and heavy elements such as plutonium and uranium come out of the oven like loaves of Wonder Bread. Not only are these elements created, but they are also ejected into the vast expanse of space due to the extreme explosion that we call supernova.

Besides the heavy elements, single particles, such as protons, alpha particles, and heavier ions, zoom out with more energy than I get from a triple-shot latte. It is estimated that they can acquire PeV ($10^{15}$ eV), or more. But the higher-energy cosmic rays are still mysterious.

As measurements of cosmic rays became more abundant, it was generally agreed that they interacted with magnetic fields, which proves that they must be charged particles. In 1938 Pierre Auger climbed high into the Alps, placing particle detectors several meters apart. The world is full of strange things, and the mountains are no exception: Both detectors registered particles at the same time.

> **?**
>
> **Question:**
>
> *Why did two independent detectors placed far apart (lateral displacement, like two different rooms on the same floor) register particles at the same time?*

- ☐ 1. Coincidence.
- ☐ 2. Cosmic rays come in pairs.
- ☐ 3. Auger was too busy skiing, and was not careful.
- ☐ 4. It was really just one particle, but, when it smashed into a molecule higher up, it created a bunch of particles, like a shower.

The other day I was attending a conference at the Inner Harbor in Baltimore when I was locked out of my hotel room. I went to the front desk and told the receptionist my tale of woe, who unhesitatingly sent my cardkey through some clunky contraption and said I was fine. My health was never in question, but, just as the manager walked by, I politely asked why things went afoul.

"It must have been a glitch," says the receptionist.

"Oh, was that the problem?" I ask.

"Yes, it was a glitch," says the manager, a little louder.

"It had me stumped," I say. "You think it was a glitch?"

"Oh yeah, it was a glitch."

They both nod and apologize for the glitch. This little parable is my way of saying number 2 (as is number 1) is the wrong answer: It does not tell us anything. I suppose Pierre may have done some skiing, but he was careful, which leaves 4 as the answer. In fact, the "shower" particles can hit other particles and make more showers, and so on.

If you look at the table on page 70, you will see that 1 GeV is the mass energy of a proton, and about the mass energy of a neutron. What I mean is that, in a reaction, a neutron can disappear, leaving behind 1 GeV (938 MeV is more accurate) of energy, as photons or kinetic energy. A 1 TeV particle has enough energy to create 1,000 neutrons, so you can see that a 1 PeV particle can create a true shower of particles. This also explains how there can be showers of showers of showers. It has been estimated that 100 billion particles can strike the Earth from a single cosmic ray.

Often enough I read something I cannot improve upon, and here is one such quote:

> The Earth's atmosphere is continuously bombarded by relativistic particles that have a kinetic energy equivalent to that of a tennis ball moving at 100 km/h. Such particles strike every 100 $km^2$ of the Earth's surface about once a year. They form the tail of the cosmic-ray spectrum, which extends from 1 GeV to beyond $10^{20}$ eV. Because of their rarity we know relatively little about them; in particular, we do not understand how or where these particles gain their remarkable energies. They represent matter in the most extreme departure from thermal equilibrium

found anywhere in the universe and may be evidence of unknown physics or of exotic particles formed in the early universe. They are possibly the only samples of extragalactic material that we can detect directly.[2]

(Of course the authors are ignoring photons in their last sentence, and the little people who work on crop circles.) With this as their opening paragraph, who could resist reading such an article? Not me. It was written in 2000, but is not out of date. These authors call cosmic rays that have an energy greater than an EeV UHECRs, for Ultra-High Energy Cosmic Rays, and I will adopt this initialism.

We can account for cosmic rays with energies up to about one PeV by supernova explosions, as I explained. It is possible for PeV particles to interact with other supernova remnants and get boosted up to about an EeV, but there is no way to prove this. As the energy continues to increase above an EeV, it becomes more and more difficult to concoct an explanation for their existence. But there is another problem that makes all this even more difficult to understand...

Charles Dickens opened his famous novel with, "It was the best of times, it was the worst of times," which might apply to our current contretemps, which is the tale of two problems. We do not know how the highest-energy cosmic rays acquire their energy, and now I will mention the other, perhaps deeper problem: It is impossible for the highest-energy particles to reach the Earth!

To understand this we have to step back to the year the first commercial nuclear power went online, the year Margaret Smith won Wimbledon in straight sets, and the year President John F. Kennedy was assassinated. It was 1963, and Bell Labs in Holmdel, New Jersey, was having trouble with their communications satellites. A noise was interfering with their signals, and they could not figure out from where it was coming. Arno Penzias and Robert Wilson got the job: Find out the origin of the noise.

No matter where they pointed the antenna, they received the same noise—it was coming from everywhere. For a year they checked and rechecked, built and rebuilt equipment, but still the noise came, no matter where they pointed the great antenna. Formulating a phrase that lives on today, they even suspected the radiation came from a "white dielectric material," called "bird poop" by most of us.

In Chapter 1 I said, "And so it came to pass that there were two camps, the expanding-universe groupies and the steady-state-cosmology fans...subsequent measurements prove we live in an expanding universe." The Holmdel Happenstance, once the bird poop was cleared up, was the single biggest proof of the big bang. Let me explain.

About 14 billion years ago the universe was no bigger than an apple. Even though all of the mass and energy we now observe (and presumably the mass we cannot observe) was in there, there was little structure. And it was hotter than a poker in a coal fire. So much hotter that this is a weak analogy, but I use it because I am hoping to remind you of the hot poker we discussed in Chapter 1 during our discourse on the light from hydgrogen. The universe made a great blackbody, and from then until now it has had the signature blackbody spectrum. Because the universe cools as it expands, the blackbody radiation changes, but is still with us. Today, the energy spectrum peaks at one thousandth of an eV, and it is sometimes called *microwave background radiation*. Like a balloon filled with air molecules, our universe is filled with photons of blackbody radiation (in addition to all the other photons made by stars, flashlights, and so on). And so, when I was learning the five postulates of Euclid in Nutley, New Jersey, Penzias and Wilson discovered this blackbody radiation in Holmdel, 40 miles south.

Actually, there is a little more to the story. Their measurements did not prove it was blackbody radiation. In order to do that, physicists had to measure the energy at different wavelengths and make a plot that agrees with the theoretical graph Planck had derived years before. I remember a colloquium when I was a graduate student in the 1970s. Everyone was excited about the newest measurements that prove the energy is really blackbody. The speaker was also excited, and proudly flashed his data through the overhead projector. (I must explain one more thing before I finish that story: error bars. If you take a ruler and measure shoe sizes you might write: Shirley—18 centimeters. Physicists would say: Shirley—18 centimeters, plus or minus 0.5 centimeters. The "plus or minus 0.5 centimeters" is the error bar.) The speaker flashed three points on the screen with error bars the size of Rhode Island. Any curve could have been drawn through those points, but we all wanted to believe it was blackbody. Eventually, of course, many, many points were made, and the error bars shrank, but that is usually how measurements go.

This detour into microwave background radiation was necessary because, as cosmic rays cruise the universe, they will bang into these photons. In fact, Greissen, Kuzmin, and Zatsepin realized this right away and calculated what would happen to UHECRs as they fought their way through this sea of photons. They found that if the UHECRs came from a distance of more than 50 Mpc or so, they cannot have energy above 60 EeV. They lose energy whacking into all of those blackbody photons.

Here is the problem: We detect UHECRs that have this much energy, but we do not know of astrophysical sources within that distance that can create such energetic particles. This is called the *GZK paradox*, and it is one of the biggest problems with which we are wrestling. This is what made me think of Dickens a few paragraphs ago: It is the best of times because we are on the brink of new discoveries. Experimentalists are challenged to build bigger and better detectors to measure cosmic rays, and theorists are charged with finding an explanation of both the origin of these particles and resolving the GZK paradox. It is the worst of times because we cannot understand nature, and nothing is worse than that.

There is a fascinating possible explanation, but I will let you ponder it for moment.

 **Question:**

*How can we explain the GZK paradox (in other words, how do particles above 60 EeV reach the Earth if we know they must travel more than 50 Mpc, and in so doing lose energy to the blackbody photons, limiting their energy to 60 EeV)?*

☐ 1. Special relativity (Einstein's theory that gives us $E = mc^2$) is wrong. Particles with this much energy avoid the GZK cutoff.

☐ 2. There is a new, special kind of particle that does not interact (or interacts weakly) with the background radiation.

☐ 3. There are nearby, extremely massive particles, perhaps remnants of the big bang, that decay (explode apart), producing the UHECRs we observe.

This is a hard question. I do not know the answer, and neither does anyone else. The first two answers have not garnered widespread support, but the third has generated a lot of interest. Let me explain.

Earlier I mentioned that a free neutron can instantaneously explode, giving off a proton, an electron, and an antineutrino.

(This process is called *neutron decay*, but that has always bothered me. A year ago I brought an apple for lunch, but it got hidden by piles of articles. A few months ago I found it, withered and looking more like a prune. This is what the word *decay* reminds me of—a year-old moldering apple—but the neutron goes off like a firecracker, fast and dramatically.) If the neutron had more mass, then the daughter proton would have more energy. By this line of reasoning, if there are extremely massive particles lurking throughout the universe, they may decay into particles we see as UHECRs. They may be everywhere, which includes being close, so this idea solves both issues at once. It explains how particles can have such enormous energies (they are born with it), and because they may be nearby, they avoid the GZK paradox.

I like this explanation but suspect it is not the true answer. It is a little akin to asking *where does rubber come from*, and being told it comes from the rubber tree. It helps, but the answer creates a lot of new questions, such as: What are these mother particles? From where did they come? Do they have anything to do with dark matter? What is their abundance? What is their half-life? There is an empirical formula that shows the number of cosmic rays that hit the Earth per day per square meter. The higher the energy, the less likely is a hit. Whatever model we come up with should be able to reproduce this curve—the heavy particle theory does not do this.

Of course one could argue the other way: Adjust the density and the half-lives of the overweight strangers such that it reproduces the curve. In this fashion we can predict the existence of a new distribution of matter. You see how wonderful physics can be. I have said, *With better instruments come better observations, and with better observations we begin to understand the enigma that is our universe.* So why did Argentina, Australia, Brazil, the Czech Republic, France, Germany, Italy, Mexico, the Netherlands, Poland, Portugal, Slovenia, Spain, the United Kingdom, and the United States cough up $50 million to the building of 1,600 3,000-gallon water tanks in Pampa Amarilla, in the western region of Argentina? That's enough water to wash my car until the sun turns into a red giant.

Here is an easy question.

 **Question:**
*What is this all about?*

- ❏ 1. They are building the Pierre Auger Observatory to measure cosmic rays.
- ❏ 2. They are building the Robert Millikan Observatory to measure cosmic rays.
- ❏ 3. They are building a car wash.

If you remember Auger's trek into the Alps you will know that number 1 is the correct answer (see *www.auger.org/contact/agencies.html*). As the particles streak through the atmosphere and into the lightless tank of black water, they will be moving faster than light travels in water. This means they will emit Cherenkov radiation—light that will be detected on the sides of the great tanks. (By the way, anybody who has seen a boat go by has seen Cherenkov radiation. Let me explain: If your motor conks out and you are bobbing around in the water, you can watch waves propagate away at, let us say, 1 m/s. If you are handy with motors and get it started again, you can zoom off at, say, 5 m/s, and will observe the familiar bow wave. You will not see this until you go above 1 m/s, the speed of waves in the water. In other words, if you travel above the speed of the waves in the medium, you emit a cone-shaped radiation pattern. Maine lobstermen call it the *bow wave*; physicists call it *Cherenkov radiation*.)

Before the cosmic rays take their water bath, they first buzz through a few miles of nitrogen, causing the nitrogen to fluoresce a bit. The observatory will measure this also, giving a dual capability for measuring cosmic rays. I can barely wait to see the data, but once again we are sitting in the jury room, locked in deliberations and trying to avoid a mistrial. The accused is physics as we know it, and we must decide if it is guilty of libel, by telling us erroneous theories. If it is not guilty, we must figure out the knotty conundrum of cosmic rays: What are they, how do they acquire such Herculean energy, and how do they avoid the GZK limit?

## Technical Notes

### Victor Hess Used an Electroscope

This is a simple apparatus, consisting of two thin metallic sheets touching along an edge and allowed to hang down together. If you take a book and hold it so that the binding is on top, the pages hang down together like the sheets in the electroscope. If a charge is added to the metal it will spread through the sheets. Because like charges repel, the metal sheets will push apart.

### Mass to Energy

A convenient unit of mass is the atomic mass unit, or amu: 1 amu $= 1.66054 \times 10^{-27}$ kg. The mass of a proton is 1.007825 amu, and the neutron is 1.008665012 amu. Adding these we find that the mass of two protons plus two nucleons is 4.03298 amu. This is precisely what the helium nucleus is made of—two protons and two neutrons—but the mass of the helium is 4.002603 amu—0.0304 amu *less* than its parts. I weigh 180 pounds and my wife weighs 120 pounds, so if we got on the scale together it should read 300. Imagine how confused we would be if it read 275 pounds instead. As far as nuclei go, the missing mass is converted into energy.

Let's crunch the numbers.

The speed of light is $3 \times 10^8$ m/s,

and 0.0304 amu $= 5.044 \times 10^{-19}$ kg.

This may sound like small beer, but watch:

$$E = mc^2$$
$$E = 5.044 \times 10^{-19} \times (3 \times 10^8)^2$$
$$E = 4.54 \times 10^{-12}$$

...which is in joules, because I was working in SI units, but if we convert to eV we get 28 MeV, which is quite a punch. For every helium atom that is formed, the sun gives off 28 MeV. If you make a measly 1 kg of helium ($1.5 \times 10^{26}$ atoms) you get about $7 \times 10^{14}$ J, which is about 163 megatons of TNT, nearly 15 times the energy released by Little Boy, the atomic bomb the United States unleashed on Hiroshima.

**Fusion Versus Fission**

The difference is a lot more than an extra $s$.

Imagine holding a magnet 1 foot over a loose (iron) paperclip on the table. The paper clip stays on the table because the gravitational force down is stronger than the upward magnetic force. Bring the magnet closer, to within an inch or so, and the paper clip flies up. Now the magnetic force is stronger than the gravitational force. It is similar for the nuclear attractive force and the electric repulsive force for two protons. The attractive nuclear force is very short-range, and dies out as the distance gets large. For something like uranium, the electric repulsion acts across the entire nucleus, whereas the nuclear attraction only acts among the neutrons and protons that are touching, or very close.

*– Chapter Four –*

# Renormalization

The 1920s gave birth to the biggest revolution in physics. The new regime was so difficult to live under that some physicists, including Einstein, could not fully abide by the new order. This new establishment broke the old totalitarianism and brought us into a world in which our knowledge is forever limited, a world stranger than any science fiction writer could have imagined. It is a world in which a particle can be in two different places at the same time, yet a world in which we can we never know its exact location. It is a world in which we cannot measure something without changing it, and it is a world in which determinism abrogates to probability—it is the reign of quantum mechanics (QM). To fully appreciate its significance, we should take a quantum leap backward and understand how physicists viewed things before QM.

Not to be outdone by the great composers, we call this previous era *classical physics*. A good place to start is with the English physicist Michael Faraday, who lived in the days when you could buy some wires and go in your basement and conduct important experiments. That was the early to mid-19th century, when Faraday discovered, among many other things, that he could create an electric field by wiggling a magnet. He derived some of the basic laws of electricity and magnetism, and is one of the fathers of this discipline.

As an example, consider a positive charge, say, the nucleus of an atom. It attracts electrons that can be relatively far away. How? Faraday gave us the notion of a field that permeates all space, and field lines that emanate out from the charge. These field lines are not real, as is a pound of uncooked spaghetti, but are a mental picture we use. These incorporeal fingers reach out from the nucleus, and, in touching the electron, try and pull it in. The same thing holds for Newton's theory of gravity, in which the forces are also represented by fields and lines of force.

These fields are determined by solving equations that describe the charges, or the masses. In addition, we have Newton's law of motion, which, once we know the fields, tells us everything about the particle. It tells us its position, velocity, acceleration, momentum, energy—everything. Also, it tells us these values exactly. This is a crucial point: According to classical physics, we can know everything, and we can know it exactly.

Here is a concrete example: Suppose we solve Newton's laws for the problem of a planet circling the sun. When we solve for the position as a function of time, we see that the orbit is an ellipse. Once we measure the position and speed at one point, we know exactly where it will be for all time.

Let us visit those times when Proctor and Gamble was learning how to make Ivory soap float and Henry Ford built his first car. You could buy the recently published *The Strange Case of Dr. Jekyll and Mr. Hyde* by Robert Louis Stevenson, and even read it at night, with electric lights powered by Westinghouse generators at Niagara Falls. It was the 1890s, and as raucous celebrations were being planned for the end of the century, some people began thinking physics was pretty much wrapped up. The few remaining problems were the last flies of autumn, about to be swept away with the broom of industry.

However, there were a few little conundrums.

**? Question:**

*What is an example of an unexplained problem that dates from around the turn of the century?*

- ❏ 1. Mystery rays coming from uranium.
- ❏ 2. The spectrum of hydrogen and other elements.
- ❏ 3. A theory that describes the radiation from a glowing, red-hot poker.
- ❏ 4. All of the above.

Once again the answer is 4, and let me go back to Rutherford, who used the alpha particles from uranium to bombard a thin film of gold. Actually, he was doing a lot of things with this new form of energy, but in order to properly characterize it, he designed this control experiment, blasting a wispy layer of gold with alpha particles. Back then, everybody knew that atoms were a few nanometers in diameter. The positive charge was smeared throughout the center like plum pudding, with electrons jiggling on the surface like tiny raisins. Armed with this knowledge, Rutherford calculated that as the alpha particles went through the foil, they should be deflected, on average, about 1 degree.

You can imagine his surprise when some were deflected by 90 degrees or more—but you should not be surprised by a question.

**Question:**

*What caused these inexplicably large deflections?*

☐ 1. Rutherford goofed in his calculations.

☐ 2. His two assistants, Geiger and Marsden, sabotaged the experiment.

☐ 3. The alpha rays were accompanied by N rays, a new form of energy.

☐ 4. The model of the nucleus was wrong, which threw off his calculations.

The correct answer is (drum roll) 4. In fact, this was one of the greatest discoveries of the century. Rutherford tried to think about what could make the force large enough to deflect the alpha particle by such an enormous amount. He hit upon the idea that, if the charge were concentrated in a tiny region, then, if every once in a while an alpha particle got very close, the force would be very large, accounting for the rare but significant instances of large scattering angles. But how small? Rutherford calculated that the nucleus (although he called it the *charge center*) would have to be a millionth of a nanometer, give or take.

On a personal note, I think this is a great way to do physics. Scratch out some calculations, do some experiments, put your feet up and think, do some more calculations, and then create a revolution in how we see the universe. But, back to the story...as is often the case, solving one mystery only triggers an avalanche that buries us in new unsolved problems.

The new problem? We now have an atom with a positively charged nucleus and negative electrons a stone's throw away. The force pulling them together is stronger than the smell of week-old fish, so what keeps them apart? One bad idea is that an atom is like a tiny solar system, with the electrons whirring around like nanoplanets. The problem is that the electrons would radiate energy, just as they do in a radio tower, which would force them into the nucleus in less than a nanosecond. Every atom in the universe would collapse before you could blink an eye—as I said, it is a bad idea. (If you read Chapter 1 you will remember I said that, on the atomic scale, Newton's laws crash and burn like the Tunguska comet. This is what I meant. The laws of classical physics fail.)

In addition to this brain buster, no one could understand why the atoms emit light in the discrete spectra I showed you in Chapter 1. For hydrogen, Swiss teacher Johann Balmer figured out a mathematical formula that predicted the wavelength of the emitted light, but no one understood the physics of it.

And then there was the ultraviolet catastrophe. This had all the great theoreticians scratching their noggins, especially those interested in thermodynamics. Everyone understood that an object, such as a hot lump of coal, radiates energy. To make things precise, and to remove the particular signature of a given object, they began to study blackbodies. These are closed, hollow metal boxes with a tiny peephole. They would heat them up and measure the radiation—blackbody radiation—that comes out of the hole, breaking it into a rainbow-like spectrum with a prism (or diffraction grating). Everyone who performed the experiment measured the same thing. Very nice. Very nice, except for the fact that when they used theory to predict how much energy was being radiated, they calculated infinity. Con Edison would love to tap into such a source, but everyone knew the theory was wrong. Theory worked okay for the long wavelengths, but, instead of predicting that the energy falls off in the ultraviolet region, it predicted that it became infinite—a true catastrophe, theoretically.

The solution to these questions would revolutionize physics and forever change our view of the world. I cannot help repeating my thematic question: Will answers to the questions I am explaining in this book bring about a new revolution so dramatic that our best physicists will refuse to accept the new world? I do not know, but let us get on with our little story anyway.

Max Planck took a great step when he invented the photon, a quantum (or discrete piece) of light. When Planck described blackbody radiation by assuming that the energy came in tiny discrete lumps (photons), he was able to prove the correct theory of the radiation, and his theoretical curve exactly matched the experimental data. (It is similar to a staircase versus a ramp, and to get the total energy you must count all of the steps. Planck broke the continuous spectrum, the ramp, into small but discrete steps—the photons. When you count the stairs you get 12 or 14, but the ramp has infinitely many tiny stairs, and when you count them you get infinity.)

Thus was born the quantum, but physics really gave birth to twins. With the *idea* of the quantum came the *number* we now call Planck's constant: $h$. I give you some numbers in the technical notes, but basically, when $h$ is multiplied by the frequency of light, you get the energy of the photon—the smallest piece of light energy. When you study atomic physics, and find that the angular momentum is quantized, it comes in multiples of $h$, Planck's constant. So, energy, momentum—everything we measure—is quantized, and comes in discrete amounts proportional to $h$. This has an even deeper significance when we try to understand quantum gravity in the coming chapters.

But what are the mechanics of the quantum? If you commute the words you have *quantum mechanics*, which governs all things small. Let me compare quantum mechanics (QM) to classical mechanics (CM) for a particle: In CM we have a function, $r$, which tells the position of the particle. In QM we have a function, $\Psi$, (pronounced "psi," although some skip the $p$), which contains all of the information we can possibly know about the particle. But it does not tell us everything. In CM, from $r$, we can find the velocity, the momentum, the energy—anything we can measure. We can find these quantities exactly and at the same time. From $\Psi$ we can only find the *probability* of a particle being in some location; we can never know exactly where it is. The only exception to this occurs when, if we know something exactly, we know absolutely nothing about another observable quantity. For example, if we know the exact momentum of a particle, we have no idea where it is!

According to CM, the particle can have any energy. According to QM, it is quantized, coming in discrete levels. This explains the atomic spectra we saw in Chapter 1, by the way. The energy levels are quantized, and, when the atom undergoes a transition from one level to another, it emits (or absorbs) a photon. In a heartbeat, the atom emits more photons than the number of stars in a galaxy, and these act together as a wave.

According to CM, the energy of a particle can be zero. According to QM, the energy of a particle *cannot* be zero. According to CM, if we know the position and momentum of a particle at any time, we know its position for all future times. According to QM, we cannot even know the position and momentum at the same time. In CM, energy and momentum are conserved. In QM, for short times, these are not conserved. Particles that, for a fleeting moment, violate these cherished conservation laws, are called *virtual particles*.

When it comes to fields, CM and QM are equally disparate. CM uses lines of force to describe fields; QM describes this by a collection of virtual particles. The energy contained in a field in CM can be anything; the energy of a QM field is infinite. Is this a misprint? After all, we cannot go around with infinite energy. The calculations predict that the lowest energy is infinite, where it should be zero, so we call this the zero-point energy.

**? Question:**

*What do we do about the zero-point energy?*

- ❑ 1. Forget about it. (I was raised in New Jersey and heard this phrase more often than my name.)
- ❑ 2. Ignore it.
- ❑ 3. Throw it away.

❏ 4. If you think about it, we only measure the difference in energy, so the overall value, even if it is infinity, does not matter.

❏ 5. All of the above.

Number 4 justifies 1 through 3, so number 5 is the correct answer.

When I was a graduate student, my advisor was William McKinley, who smoked a pipe (all the physics professors did back then). His office was smokier than my kitchen, but every once in a while he would stop puffing and smash the pipe on the inside of his trash can, sending glowing embers in the air like lava pellets from Vesuvius. He seemed to do this much more frequently when I was saying something stupid, so I always tried to get my physics right before I ventured into the inferno.

One day I was explaining my theory of the mass spectrum of elementary particles and he started to pound like a blacksmith. Feeling sympathy for his neighbors I backed down from my theory, but because I was interested in quantum field theory and general relativity, he suggested I look into the relation between the two.

My first surprise came when I found out that the zero-point energy of the electromagnetic field had been measured. All the textbooks I was reading invoked the directive, "normal order the Lagranigan," which is a fancy way of saying you throw the zero-point energy away. All of a sudden I am reading how Casimir predicted that, due to the zero-point energy, there should be an attraction between two neutral metal plates. The calculations were long but not too difficult, and I agreed with his result. This force is weaker than the power of Congress, but it was measured in Phillips' lab in 1948.

In the next decade Tim Boyer redid the calculations for two concentric spheres. If you picture this, any small region is similar to two parallel plates. For example, imagine one sphere 1,000 mm in diameter, nestled inside a sphere of diameter 1,001 mm. Now stick a small postage stamp on the sphere. For that region, you can hardly see the curvature, and the two shells, locally, look like two parallel plates. Following the previous result, it certainly seems that there should be an attraction between the two plates. (This is how physicists think about things, by the way: Use a simple argument to guess the answer, and then do the often long and difficult calculations to prove it.) Boyer showed that, indeed, there is a small force due to the zero-point energy, but it is repulsive!

His calculations went on for pages and pages; it was the most difficult calculation I had ever seen, and I did not re-derive every step—a no-no for theoretical physicists. At the same time, my advisor was telling me to explain all this. My research took me into general relativity, and I will come back to that later, but I will tell you this now: The result stands today, and everybody still shakes their heads over it.

My point in this digression is to elucidate that QM is very different from CM. But there is no doubt that QM is correct. The spectrum of hydrogen, and all the other elements, are correctly predicted by QM. Transistors, light-emitting diodes, laser diodes, gas lasers—are all described by QM. Alpha radiation, beta radiation and gamma radiation, nuclear radioactivity, fission, and fusion are all described by QM. All of the scattering experiments from the particle accelerators from CERN (the European Organization for Nuclear Research) to Stanford, the discovery that neutrons and protons are made up from smaller particles (quarks), are all described by QM. The strong bonds that hold atoms together making molecules, and the weak bonds that can hold noble gases together, are explained by QM. Without QM we would not understand why different objects have different colors; without QM we would not understand what holds us up; without QM we would not understand why the sky is blue. Without QM we would not understand anything. (Sometimes I exaggerate a tad.)

Now I will teach you some quantum field theory. A good way to start is with the electromagnetic field. In the past, we pictured this as a continuous field, as Faraday did. This is a very passive view, like prairie grass undulating to the wind. In quantum field theory the field comes alive, taking on a much more active role than wimpy blades of grass. Photons are created and annihilated left and right, all the time. It is through the exchange of these virtual particles that real particles interact.

Let me give you an analogy: Caravan C is traveling in a straight line across golden sands of the searing desert. It has scouts that dart out and back, scurrying about, checking for water, weather, and bad guys. Caravan X is doing the same thing. Suppose a scout from C goes to X, telling them where he is from, and a scout from X is sent to C, telling them of water X just passed. Caravan C then alters course toward the water, and, after C and X are once again separated, the scouts can no longer interact, and the caravans go in straight lines.

Particles interact in the same way. Do not think of an electron as a porcupine with long quills (Faraday field lines); think of it as an old dog with fleas continually popping off and landing. An electron travels surrounded by this cloud of virtual photons. When another electron goes by, one or more of these photons are exchanged, and the electrons alter their course. The exchange of these apparatchiks is how we account for the force between particles.

There is one additional fact about these exchange particles I must mention: They can be massive or massless. I talked about massless particles in Chapter 2, but the important thing here is that massive exchange particles give rise to very short-range forces. An example is the nuclear force, which dies out as soon as a particle gets more than a nuclear diameter or two away. Massless particles give range to long-range forces such as the electric field.

Mathematically we account for the creation and annihilation of particles in a remarkably simple way: We have creation and annihilation operators. We replace the continuous field with a collection of creation and annihilation operators. Now—bear with me—we envision a collection of all possible states the system may be in, and we call this collection a Fock space. Creation and annihilation operators work on this Fock space, changing the state from one to another.

Let me give you a more concrete analogy: Imagine a fly and a gnat buzzing in and out of a three-room house. A classical description tracks them through the kitchen, guest room, and bath, knowing their exact location with arbitrarily high accuracy. In quantum mechanics we create a Fock space of all possible states: Both bugs in the kitchen is one state; both bugs in the guest room is another state; the gnat in the kitchen and the fly in the bath is another state; the fly in the kitchen and the gnat in the bath is another state; and so on. There are also the empty-house and the one-bug states. There are 16 states in all. Let us denote the states, in the order I just gave them, this way: $(FG,0,0)$, $(0,FG,0)$, $(G,0,F)$, $(F,0,G)$, and $(0,0,0)$. Let $a^+_{Fk}$ be an operator that creates a fly in the kitchen: $a^+_{Fk}(0,0,0) = (F,0,0)$. In the same vein, $a^+_{Gb}$ creates a gnat in the bathroom. The annihilation operators look similar, but have no $+$, so $a_{Fg}$ annihilates a fly from the guest room: $a_{Fg}(G,F,0) = (G,0,0)$. Armed with the Fock space and the creation and annihilation operators, we can account for all the properties that can be measured. If I go into any more detail you might end up becoming a physicist, so I'd better stop.

I think you can see the difference between the classical and the quantum. Our intuition sides with the classical view, as we envision the fly swoop by a window and zigzag through a doorway. But at the atomic level, where the bugs are particles, the pointillistic creation-annihilation approach is the only way we can describe nature. I want to make sure you are with me, so I will ask you a question. If you get it right, keep reading. If you get it wrong, you must reread the previous paragraphs.

**Question:**

*Which of the following are classical concepts, and which are quantum concepts?*

☐ 1. The magnetic field of the Earth can be determined exactly, with absolute precision.

☐ 2. The electric field from a nucleus is viewed in terms of field lines, which are continuous.

☐ 3. When two particles interact, they exchange particles.

☐ 4. In general, we cannot determine observable quantities (such as momentum or position) with unlimited accuracy. There is always some uncertainty.

If you thought that the first two are classical concepts, and the last two come from our quantum view of nature, keep reading.

One last point (which is not really my last point): Which description is correct: quantum or classical? The answer is quantum, but there are limits where the classical description is close enough. This is embodied in the correspondence principle, which states that in the appropriate limit, quantum turns into classical. For example, imagine a photon cruising toward a plate-glass window. This is high-quality, clear glass, so the photon is not absorbed, so what happens to it? The answer is: nobody knows. If you send 1,000 photons, I can tell you that about 40 will be reflected and the rest transmitted. (I am assuming normal incidence.) In fact, QM should predict that the probability of reflection is 4 percent, but I have never done this calculation. On the other hand, if you assume you have zillions of photons, you simply assume we have a classical wave, and every physics major in their sophomore or junior year can derive that the reflection is 4 percent. Many photons act like a wave, quantum to classical.

In summary, everything is perfect. Quantum rules, and we even understand how the classical approximation arises. Time to sit back and rake in the profits from telephones that take pictures and automatic cars that know how to park. Except for two things: gravity and infinity. Gravity cannot be quantized, which I will discuss in the next chapter. The other issue is infinity, which I will tackle now.

Most of the time we cannot solve anything exactly. Therefore, we resort to approximation and perturbation schemes. For example, when you consider that the sun has nine (oops, I should have said eight) planets, and most of those planets have moons, you have a system far to complicated to solve exactly. So what we do is start with the two-body problem, which we can solve exactly. We solve the idealized problem of the Earth revolving around the sun, ignoring everything else. We do this for each planet, calling it the zero-order approximation. Now we can calculate the effect (perturbation) Jupiter has on Earth, by using the solutions we just found, and so on. The reason this works is that the perturbations are small. Virtually every measurement we make is compared to calculations that arise from the perturbation approach, often many perturbations. Each additional perturbation brings us closer to the exact answer.

Let us revisit hydrogen and the four visible lines you saw in the first chapter. As diffraction gratings improved, the spectrum could be viewed much more carefully: It is similar to looking at your skin with a magnifying glass, and then switching to a microscope. The famous four lines were seen to be split apart, signaling new physics at work. Let us make a quick visit to this exciting decade of the 20th century, when the terrible carnage of the Second World War ended and the most eminent physicists were searching for new directions.

It was 1947, Shelter Island, and top brains gathered to find postwar directions for physics. One of the invitees was Willis Lamb, who had done an extraordinary experiment. He found, using the most careful measurements the times allowed, that one of the lines in hydrogen was split. It was a tiny, tiny separation, but the theory said there should be one line, and Lamb found two, hugging like star-crossed lovers. It was known that some lines split because of the magnetic field of the particles, but the Lamb shift resided where all theory said there should be none. This is another example of: *With better instruments come better observations, and with better observations we begin to understand the enigma that is our universe.*

This effect was not nearly as dramatic as the extraordinarily large deflections Rutherford saw; it was more like pesky Mercury, where things were just a little off (those 43 seconds of an arc per century). The solution to Mercury's orbit revolutionized our view of the world, bringing us from Newton's absolute three-dimensional space to Einstein's willowy four-dimensional space-time. Could Lamb's shift be as significant? The physicists at Shelter Island wanted to know.

When I was a graduate student I studied nuclear physics with Joe Levinger, who was a student of Hans Bethe. Bethe was one of the greatest American physicists, and Joe always had a good story to tell about Bethe. The most pertinent one was how he calculated the shift that Lamb measured, giving it a theoretical foundation. Bethe used the old dog with fleas, assuming that the electric field was quantized, consisting of photons. Using quantum field theory, Bethe calculated the Lamb shift, and when I redid the calculation 30 years later as a graduate student, I thought my head would explode.

The perturbation method I described a moment ago is used, but a problem crops up. The first-order approximation works fine, but the second order, which should give an even smaller correction, gives infinity! Bethe was so smart he was able to get (very close to) the right answer by sidestepping the hulking monster that was infinity, but, in general, physics was in big trouble.

Virtually all higher-order corrections gave infinity, and the whole theory hovered at the edge of the sink, waiting to be washed down the drain. But physicists are clever, and the tap was flowing with ideas on how to salvage physics. The lifesaver was renormalization.

The idea behind renormalization is, first, to admit there is a mistake somewhere. Consider our picture of an electron of mass $m$, the dog with fleas. Instead of using the bare mass, just the dog, we should consider the physical mass, dog with fleas. In other words, the mass of an electron should be the "bare mass," plus the mass energy of all those virtual particles. Because the calculations give us infinity, we should allow that this "dressed mass" can be infinity too. Now we can go back and redo the calculations and balance the infinite mass with the previous infinite result, leaving behind a small number that represents what we measure.

This may sound homemade, but there are two reasons why it became accepted. First, the small number left behind gives the right answer (or very close to it), and second, you only renormalize once. As you go to higher and higher orders in the perturbations, they are automatically finite, and, even better, accurate.

But infinity is stranger than a book without ink, and very few physicists are comfortable with this procedure, which is why I call it a benign tumor. We can live with it, but we would rather have a theory without such a thing in its heart. Let me give you a famous example that shows how strange infinity is. Suppose you have a hotel with a person in each room. It is full and cannot accommodate another guest. However, if the full hotel has infinitely many rooms, than you can accommodate another guest. Simply move the guest in room 1 to room 2, the guest from room 2 to room 3, and so on, and put the new traveler in room 1. In fact, it can be proved that this full hotel can accommodate an infinite number of additional guests.[1]

Another example: Consider all of the positive integers, 1, 2, 3, .... There are infinitely many of them. Now consider all of the even integers, 2, 4, 6, .... There are infinitely many of them also. Now, you might think there are more integers than even integers. For example, in the range 1–100 there are 100 integers, but only 50 even integers. You can do this for any range you want, and you might even conclude that there are twice as many integers as even integers, but you would be wrong. There are just as many even integers as integers. Let me give one more example because I cannot resist. Imagine all numbers, or points, on the number line from 0 to 1, which we call the unit interval. There are infinitely many points, but there are more of these than the number of integers! So, we see, there are different values of infinity. Now suppose you consider the number of points in the unit square, the edge of which is the unit interval. Which has more points: the line or the solid square? They have the same number of points, which proves my point—infinity is strange.

The usual laws of arithmetic can fail, and this business of renormalization floats on a precarious balance of intuition, convenience, and lack of alternatives. But, speaking of balance, I should point out another view, given by the great physicist Steven Weinberg. He is most renowned for his work on particle physics, but also wrote a book on general relativity that I used as a graduate student. (I still have it, with duct tape holding the Scotch tape together, but I digress.) My point is that Weinberg looked at renormalization as a tool to weed out bad theories: If it cannot be renormalized, it is wrong. A very practical approach, and makes the best of what I see as a bad business.

Renormalization is not often considered one of the pressing problems in physics, but it gets worse when we consider general relativity. I will come back to this in the next chapter, and show more reasons why infinite quantities are unacceptable.

## Technical Notes

### The Quantum

I wrote, "...when $h$ is multiplied by the frequency of light, you get the energy of the photon," so let's do it. Mathematically the formula is nice and simple:

$E = h\nu$, where $\nu$ is the frequency of light.

Suppose we have green light, which has a frequency of

$6 \times 10^{14}$ Hz (cycles per second).

Planck's constant is $h = 6.67 \times 10^{-34}$ Js (joule seconds), so

$E = 2.5 \times 10^{-19}$ J, which is about

2.5 eV.

### Quantum Field Theory

Sometimes we distinguish between quantum mechanics (QM) and quantum field theory (QFT). *QM* refers to a system in which the number of particles is fixed, and the fields (such as the electromagnetic field) are continuous classical entities. This is the original QM developed by Schrödinger, Heisenberg, and others. QFT, sometimes (but not much anymore) called *second quantization*, goes a step further and allows for the creation and annihilation of particles, such that the fields are no longer classical but are also described by annihilation and creation operators. Often, however, the phrase *quantum mechanics* refers to the entire notion of our quantum view, including QFT, and I have used it in this sense in this book.

*– Chapter Five –*

# The Higgs Particle

In Chapter 1 we saw the visible spectrum of atoms. As you know by now, the spectrum results because the energy levels are quantized, and, when an electron fires down from an upper to lower energy level, the atom spits out a photon. All this is explained by quantum mechanics. So let me ask you a question right away.

**Question:**

*Consider the nucleus of any atom, consisting of protons and neutrons. Does this also have quantized energy levels, and does it emit photons during a transition?*

- ❏ 1. No. It is too small.
- ❏ 2. Yes, but they are invisible.
- ❏ 3. No. Quantum mechanics only applies to atoms, not to nuclei.
- ❏ 4. Yes. That is where we get gamma rays.

This has two correct answers: 2 and 4. Think back to Paul Ulrich of Chapter 3, who discovered gamma radiation from uranium. The nucleus gives off radiation just as an atom does, but, because the nuclear force is so strong, the energy of the photons is much higher, in the MeV range, a million times more energetic than the eV-range photons of visible light.

And so the realm of nuclear physics was born, wherein physicists began to understand the nature of the emitted photons and appreciate the vast energies tucked away at the core of nature's tiny gift: the nucleus.

Picture this: I am in graduate school taking the second semester of nuclear physics as the professor scratches the board with equations and energy levels, but I do not hear a thing. Cathy Fiore is the prettiest woman in the school and she is sitting next to me. She will end up at MIT's Lincoln Lab, so I guess she is paying attention, but I am thinking neither about that nor nuclear physics. (I paid the price, by the way, getting a B in the course, which is as low as you can afford in graduate school. Luckily she was not in my other courses, so the As carried the awful B.) One thing I do remember, though, is about mirror nuclei, which are different nuclei with nearly identical energy levels. The brilliant German physicist Werner Heisenberg, co-discoverer of quantum mechanics, had an idea about these. Consider, he said, that the proton and the neutron are really different states of the same particle—the nucleon. Now, suppose you make up an operator that can change protons to neutrons and vice versa. If you view the nuclear particles as nucleons, then this operator does nothing, and the energy states should not change under the operation. Mathematically these operations can be represented by 2-by-2 matrices, which are collections of numbers with two rows and two columns.

I always try to poke my nose into my kids' homework, especially when they are doing math. One day I said to my high-schooler, "Wow, Katie—you are doing matrix multiplication. Cool."

"Boring," she responded.

A hundred years ago no one outside a small cadre of mathematicians ever heard of a matrix, but Heisenberg learned, and he used them. For the hardy I give a few details in the Technical Notes, but you see how their use has mushroomed. Not just any old 2-by-2 matrices are sufficient for the operators I just mentioned; they must be Special and Unitary (so that probability makes sense and we can include transformations that only change things by a little bit). We call these SU(2).

The main thing I want you to remember is that nuclear forces are invariant (do not change) under SU(2). This concept (and generalizations) is central to all our understanding of all of the particles we have ever observed. It makes the assumption that protons and neutrons are the two different states of the same particle, and if you change one into another, by using the SU(2) matrices, physics does not change. If you Google SU(2) you will find enough literature to fill a library.

We call this a *symmetry*. Suppose you take a chessboard, and, keeping it flat on the table, rotate it 180 degrees (not 90) about the center. It appears to be exactly the same, and we call this a *symmetry* also. We know that the proton and neutron are really different though, because one has charge and one does not, so the symmetry is not exact, and we say that the symmetry is broken.

You do not need to be a nuclear physicist to understand this: Everybody learns about broken symmetry when they play musical chairs. A line of chairs is placed facing the Master of Ceremonies. When the dark strains of Bach's fugue echo from the walls, people run all over—north, south, east, and west. This is symmetric, for if you rotate the room by 90 degrees (or any amount), people would still be buzzing around in all directions. When you stop the tunes, you break the symmetry. Everybody sits and everyone is facing east (or wherever). The symmetry is broken.

Using symmetries is an important and exciting new way of looking at nature, but there is a problem. To explain it, let us step back to 1905 for a moment. We are in Germany, and the physics community is rocked by Einstein's special theory of relativity. It seems strange and cuts against the grain of common sense, and there is little direct evidence to prove the theory. Physicists struggle to know if it is correct, and contemplate the repercussions of such a bizarre new world.

One well-known implication is that nothing can travel faster than the speed of light. This fact alone is the sword that kills Newton's theory of gravity! The reason is this: Time appears nowhere in that theory. Consider the force between the Earth and the moon. It is inversely proportional to the square of the distance between them. It depends only on distance. Now, if an angel came along and nudged the moon, the gravitational field on Earth would change a bit. In fact, the angel could continue to wiggle the moon back and forth and communicate to us using this digital signal. The change in the value of the field would be felt *instantaneously* here on Earth, but this violates special relativity. Nothing can travel faster than the speed of light, which takes more than a second to reach us from the moon.

With neither a single experiment nor the stroke of a pencil we find: if special relativity is right, Newton is wrong. Einstein was as sensitive as anyone on this point, and 10 years later would develop general relativity, a theory of gravitation that does not violate special relativity, but this is the subject of another chapter.

Back to the ideas of Heisenberg, the problem I alluded to is that the SU(2) transformation is instantaneous, or global, affecting the entire space at once. This violates special relativity as egregiously as Newton's theory of gravity. A few physicists were taking this issue very seriously, one of whom was Japanese physicist Ryoyo Utiyama. He thought about correcting this contradictory state of affairs, and was also laying a modern foundation for the theory of gravity. As Utiyama was sketching out his ideas he was invited to Princeton, but soon read a paper that made his heart sink. Someone else had just published a paper, not about gravity, but about elementary particles, that overcame this inconsistency.

It is a now famous paper by Yang and Mills, published in 1954. They changed the "global" invariance to a "local" invariance. Mathematically this means that the transformation depends on space and time. Physically it means that signals do not propagate faster than the speed of light, and introduces a new field, the local gauge field, that keeps everything invariant. Yang and Mills called this new field the **b** field, and this has come to be the heart of how physicists look at all forces. Utiyama, by the way, published his work in 1955 and laid the foundation for a local gauge theory of gravity, but there are few prizes awarded for being second.

One of the reasons the Yang-Mills paper is so famous is because it is wrong. That may sound a little harsh, but it is not meant to be an animadversion; it is meant to be a tribute to how important and compelling their ideas are. They were trying to account for the nuclear force, but the exchange particles were massless. I described in Chapter 4 that nuclear forces are short-range and require a massive exchange particle, but massive particles destroy local-gauge invariance. Yang and Mills were perfectly aware of this problem, and said, "We next come to the question of the mass of the **b** quantum, to which we do not have a satisfactory answer." You saw how much I like this sentence; it inspired a putative title for this book.

The idea was so compelling that it was published anyway, but the mass problem flew on the flag of conundrums for years, waiting for a solution. Let me sum up the impasse. Everybody liked the idea of local-gauge invariance. Local-gauge invariance required massless exchange particles. Nuclear forces involve massive exchange particles. Hmm.

While this flag is flapping in the breeze another great idea comes along. Let us think about atoms one more time. They are very different, yet all are made from the same few particles. Suppose we extend this thinking to the nucleons, and assume neutrons and protons are made from something more fundamental. These "more fundamental" particles are called quarks. There is the up quark, $u$, and the down quark, $d$, and four others. We always denote the charge of the proton by $e$, which is positive. Remember that the charge of the neutron is zero.

**Question:**

*How can you make a neutron* n *and a proton* p *from a total of three up-and-down quarks? (Hint: Assume the up quark has a charge of 2e/3 and the down quark has a charge of –e/3.*

☐ 1. You cannot; it is a silly idea.

☐ 2. $p$ = 2 ups and one down,
$n$ = 1 up and 2 downs.

☐ 3. $p$ = 3 ups and $n$ = 3 downs.

You have to add fractions, but 2 gives the correct charges of 1 for the proton and 0 for the neutron.

Every time my wife asks me to fix something I remind her that nothing is simple. I have said it so many times we do not even argue about the logic, or philosophy, of my response anymore, but it comes into play here as well. In order to use the quarks to make a local-gauge theory, each one has to come in three different states. This is similar to the nucleon, which comes in two states (the proton and the neutron). Similarly, each quark comes in three different states. We call these different states *colors*, and label them as *red*, *blue*, and *green*. The nucleon comes in two states so we need 2-by-2 matrices. The quarks have three, so we need 3-by-3 matrices, and so the symmetry is SU(3).

The quarks interact the way other particles do, by exchanging virtual particles. What are these virtual particles called? Just to show you that physicists are not devoid of humor, we call them *gluons*. Similar to the photons, they are massless and have spin 1. By the way, many physicists were reluctant to accept this. After all, they said, no one has ever observed a fractional charge (meaning less than $e$).

The answer to this involves, perhaps, the strangest thing in all of physics. Every fundamental force we have ever experienced gets weaker with distance. The farther a planet is from the sun, the weaker the field. Nuclear forces, we have seen, are very short range, and die away quickly with increasing distance. This seems reasonable, but the quark force, we believe, is the exception. Let me explain.

Suppose you want to measure this "fractional charge" by isolating a quark. Just slam something into the nucleus and look for particles that come out with fractional charge—quarks. Good idea, but we have been slamming particles together for decades and never detected a single particle with fractional charge. The explanation for such shyness is that the force between two quarks does not decrease as they get farther apart. It is like working against friction: The force never goes away. In other words, the more you pull, the more energy you put into it. With quarks, if you apply enough energy to get them 1 fermi ($10^{-15}$ meters) apart, they have enough energy to create another pair of quarks, thanks to $E = mc^2$. The more energy you put in, the more pairs of quarks you make. Thus, you will never make an isolated quark; you will never see fractional charge. Cute, no? Just to make sure you don't think this is homemade, we came up with an official name: *confinement*. Due to quark confinement, an isolated quark cannot be produced.

With the Higgs field, the Heisenberg maquette was replaced by one of the great discoveries of the 1960s, the local-gauge theory of weak interactions, the kind of force that accounts for the beta decay Rutherford found so long ago. It has a fascinating history and a surprising result—at least, it surprised the heck out of me. It goes back to Heisenberg's idea of the neutron and proton being different states of the same particle (the nucleon). Glashow, Salam, and Weinberg used this idea, but applied it to the neutrino and the electron. They assumed that these were different states of the same particle. But wait a minute, you say. This might be acceptable for the proton and neutron because they have nearly the same mass, the only essential difference being the charge, which has nothing to do with the nuclear force. But electrons and neutrinos are dogs and cats. Electrons have charge; neutrinos do not. Back then, it was believed that the neutrino was massless, nothing like a plump electron.

When I am sight-reading in music (playing at half tempo at best), sometimes things don't sound quite right. You have to "read across," meaning continue until you hear the whole idea. We must carry this leitmotif into the weak interactions, and remember that, as we see the world, the symmetries are broken. So, let go of Heisenberg's formulation and the SU(2), and apply this to the electron-neutrino. Make the gauge symmetry local, and you have the electroweak *unified* theory. One of the salient features I want to emphasize is that an isolated theory of weak interactions cannot work, because it is too protectionist, like a country with tariffs so high that all trade fails. It is the free trade between the electron and the neutrino that gives a correct theory. Also, unlike the original Yang-Mills theory, the exchange particles can now acquire mass, and these are called the W and Z bosons. In addition, of course, are the massless photons.

When all of this (and a lot more I do not discuss) is put together, we have the standard model of elementary particles. It involves $SU(3)$ (for the quarks), and $SU(2)$ (weak interaction) and $U(1)$ (for electromagnetism), so we say the standard model is the local-gauge theory of $SU(3) \times SU(2) \times U(1)$. This is a patchwork of different symmetries with as much aesthetic appeal as a copperhead. Yet it works. Although we have to put a lot in, it correctly describes the flocks of particles that graze in our labs and accelerators.

Now let me try and sum up the disparate experiments and theories that make up our view of the world. There are four forces—gravity, electromagnetic, strong nuclear, and weak nuclear. Three of them struck a deal, and like a fragile political coalition, fly under the union flag of the Standard Model. It is brokered patchwork, and, like the old USSR, each is trying to cling to its own identity, but there is one theme of common agreement: Gravity is ostracized. So let's discuss the big three for a moment. When all the ideas I mentioned come together, here is how it works: The strong nuclear force, the strongest force on a particle-to-particle basis, arises through quark interactions. A quark and antiquark can bond together to form a meson. When one nucleon interacts with another nucleon, it can exchange a meson. The weak nuclear force involves quarks, electrons, and neutrinos, and is about 13 orders of magnitude weaker than the strong force. The exchange particles are the heavy (90 GeV) W and Z bosons and acquire mass through the Higgs mechanism I will describe in a moment. The electromagnetic force is one to two orders of magnitude weaker than the strong nuclear force, and is felt across all particles with electric charge. The exchange particle is the massless photon.

For practice, think about the helium nucleus. The two protons push against each other stronger than a couple in divorce court. The two neutrons are like the marriage counselor, trying to keep them from flying off. The protons exchange photons. All of the nucleons exchange pions, holding them together. The pions are quark-antiquark pairs, held together by the gluons. The tiny little particle Rutherford named so long ago, the alpha particle, is a little kingdom bustling with commerce and industry, as complex as a multinational corporation. In Chapter 2, I mentioned that a free neutron will spontaneously explode into a proton, positron, and antineutrino. This is the weak interaction at work.

A key development in this approach was made when nuclear forces were explained by local-gauge theory. Remember the problem: The exchange particles must have mass, but local-gauge symmetry requires massless only. The solution was to take a more enlightened view concerning mass. In this way we say there really is a symmetry, but it is hidden by another field we have never seen. This hidden field permeates all of space and is as reliable as the mailman. Whenever a particle has to exhibit its mass, the symmetry is broken, and, thanks to the field, the mass appears. This is the most important field in physics. This is the Higgs field. The thing is, though, we have never observed the Higgs field. In the quantum world this field becomes particles, and we have never observed the Higgs particle either.

I never liked the word *lynchpin*, but I can think of nothing better to describe the Higgs particle. Everything we see and everything we sit on depends upon the existence of this particle, yet we have never seen it. Every massive particle in the universe owes its mass to the Higgs, but we have never observed it.

How can this be? We have searched and searched and searched, and, even though I am a theoretician, I have kept my eyes open too, but it has never been found. The most straightforward answer to not being able to find something is that it does not exist. But if the Higgs particle does not exist, our entire theory of elementary particles is wrong. This is as bad as it gets.

I have written page after page about strong forces, weak forces, exchange particles, symmetry, and broken symmetry All this leads up to the Higgs particle, the cornerstone of everything we know about elementary particles. We have found everything else we predict, including quarks, which reminds me...do you remember Rutherford's startling discovery of the nuclear atom? He deduced it from scattering alpha particles from gold. Physicists were able to do the same thing, in the 1960s and later, with nuclei. The experiments bombarded neutrons or protons with high-energy particles, and carefully measured the pattern of the scattered objects. The only correct interpretation of the data was given by assuming there were three point-like objects doing the scattering. At first, being cautious, they were called partons, but as the evidence piled higher than a New Jersey landfill, quarks took their rightful place on the stage of elementary particles.

But the Higgs particle is still offstage, hidden from direct view behind curtains as opaque as solid gold. All hope is not lost though, and, in fact, there may be a very good reason why we have not met Peters Higgs's famous particle. I will let you try your hand first.

**?**

**Question:**

*Why has not the Higgs particle been observed?*

☐ 1. Because it is neither a fermion nor a boson.
☐ 2. Because it is massless.

❑ 3. Because it was captured by the ozone.

❑ 4. Because it has too much mass, and we do not have accelerators big enough to create the energy needed to observe it.

Every particle we have ever observed and every particle we have ever predicted is either a fermion or a boson. Except, of course, for anyons. Anyons are really quasiparticles (collections of quantum states) that arise in two-dimensional systems, like a very thin slab. So they do not really count. In fact, the Higgs particle must have zero spin (for some reason physicists never say the particle has *no spin*; we say it has *zero spin* or is *spin zero*), so it is a boson. This polyonymous entity is also referred to as the *Higgs boson*. We see massless particles with every page we read, so that is not an issue, which leaves 4 as the right answer. Experimentalists and theoreticians have been working like the dickens, though, and are pretty sure its mass must be more than about 115 GeV. This is really its mass energy $mc^2$, but the "energy" is implied. If its mass were any less, it would have been observed by now. By the way, the Higgs is no featherweight: 115 GeV is about 120 proton masses.

Now, depending on your point of view, here comes the scary part or the exciting part. Based on things we have already measured, the Higgs boson cannot have a mass greater than about 150 GeV. Many physicists might say this is the most important issue in physics today: finding the Higgs boson. The Standard Model has been so successful through-out the years that I feel it will probably be observed. But an-other part of me sees great excitement if it is *not* found.

Excluding the kids who get a new Porsche for graduating, the Standard Model is like your first car: Even though it is an old clunker, you are excited because it works. And if you have trouble down the road, it can be fixed. This is the Standard Model.

I am forced to take a short digression back to when I was teaching Astronomy 110 (no, not 101, but I never under-stood course-numbering theory). Although I am a theoreti-cian, I loved to do demonstrations. The students loved it too—they quickly learned there was a good chance some-thing would backfire or blow up. I am doing the old standby, the bell-in-jar routine. There is a buzzer in a huge jar that is connected to a vacuum pump. As the air is evacuated, the sound goes away, and then, releasing a valve that lets the air back in, we all hear the sound coming back to life. It works this time, so everyone claps, and I bow, but there's a moral to the story, simple as it is: Sound needs air through which to propagate. Water waves need water; waves on ropes and slinkies need ropes and slinkies. We used to think all waves need a medium through which to propagate.

Because light was thought to be a wave, it was no excep-tion. But it turned out to be another pesky issue. We knew light came form the sun, and distant stars, so the medium must permeate the entire universe. The enormous speed of light would make this mysterious medium "stiff," yet, because the planets are unaffected by it, it must somehow be very courteous to them, offering little, if any, resistance. This odd medium was named the *aether*. Nowadays we smile at this quaint notion. Light consists of photons that travel through space more easily than they travel through air. True, in bunches they act as waves, but the aether is like rotary telephones: a thing of the past. Or is it?

Soon after the aether was pushing up daisies, Einstein postulated the cosmological constant I discussed earlier. This term fills all of space with

By the way, you can probably tell by now that physicists have had a long love affair with everything Greek. The use of quarks and gluons was a breath of fresh air into this system. But speaking of fresh air, apparently the Greek gods get better air to breathe than the rest of us. This pure air is the Aether, who is one of the "first-born." To me, understanding the Standard Model is much easier than figuring out Greek mythology, and I apologize if I do not get this aether thing quite right.

curvature. Similar to the aether, it is constant and fills our universe. In recent years the effects of a non-constant cosmological term have been reassessed, reviving the notion of a quantity that permeates all of space-time.

Now, like a plane pulling out of a nosedive, I can end this digression and ascend to the present day—the day of the Higgs. It too predicts a field that permeates all of space. Is this a true picture of nature, is this how the universe really is, or will the Higgs go the way of the aether? I cannot answer that, and neither can anybody else—yet—which is why I am writing this book.

Let us turn to some details. When all the dust settles, the Standard Model, molded from both theory and observations, describes three families of elementary particles. The first consists of the electron, the electron neutrino, and the up and down quarks. Just about all of the matter we see can be made from this family. You, me, the pooch, Pluto, and the sun are all made from these particles. But lurking beyond this is the second family, in many ways just like the first, only the particles are more massive: the muon (a heavy electron), the muon neutrino, the charmed and strange quarks. The third family is the tau (a heavy muon), the tau neutrino, and the top and bottom quarks.

When I was a graduate student, after we finished our physics we went to the Rathskeller for a few cold ones, and after that we would order a pizza. When the pizza arrived, every once in while someone would stand up and say, "Who ordered *that*?" and we all would laugh. What was so funny? Did we just have too many beers? After the muon was discovered · it was realized it is just like the electron, only more massive. In fact, it decays into the electron (plus a neutrino and antineutrino) anyway, and seemed to be an extra particle with no raison d'etre. In fact, after the discovery of the muon, particle physicist and Nobel Laureate Isidor Rabi said, "Who ordered *that*?"

Let me explain a few threats to the Standard Model. It was the 1990s and I had the best graduate student I ever had. I was working on my theory of gravity that includes both mass and spin, and the theory predicts that any fermion will have a tiny additional field, called the torsion field. I told Terry, my student, that this effect is extremely small, but it should have a slight effect that may be measurable.

After that Terry was like an avalanche, filling my boards with equations and my room with enough chalk dust to end global warming. I was the sound that triggered the movement, but after that, it was all his doing. Anyway, in the middle of all these calculations major newspapers reported that the measured value of the magnetic moment of the muon was wrong. Let me explain.

Ever since the ideas of Uhlenbeck and Goudsmit, followed by the theoretical work of Dirac, we have come to know that fermions have spin, and, if they are charged, they also produce a magnetic field. So when you think about electrons and protons, think about them as if they were tiny bar magnets, with a north pole and a south pole. We are able to make extremely sensitive measurements of the magnetic moment, the origin of the field, and in the 1990s, for the first time, theory and experiment disagreed.

I have described this situation many times in this book, and it was exciting to be in game. Was the Standard Model wrong? Were the experiments flawed? Did the theoreticians goof? Was it torsion? Was there another field?

Most of the physicists in the United States think torsion is too small to measure, so Terry and I were pretty much alone here, but for a moment Terry slowed down.

If you thought you would never see the Pythagorean Theorem again, you thought wrong; here it comes: In two-dimensional space it would read $s^2 = x^2 + y^2$, and in three dimensions it is $s^2 = x^2 + y^2 + z^2$. These hold true in Euclidian space. If you have four-dimensional Euclidian space this would become $s^2 = x^2 + y^2 + z^2 + w^2$. By the way, this little example shows you the tremendous power of mathematics. I will bet you one dollar that you cannot imagine (picture) four-dimensional space, but with this formula you can compute the distance between two points in it as easy as pie.

Hermann Minkowksi, a German mathematician, invented (discovered) space-time in 1908. He assumed:

$$s^2 = x^2 + y^2 + z^2 - (ct)^2,$$

where $t$ is time and $c$, as always, stands for the speed of light. Do you see the essential difference between Minkowski space-time and four-dimensional Euclidian space? It is the minus sign, which makes all the difference in the world. I remember reading (a translation of) Minkowski's book, *Space and Time*, and will never forget reading: "From henceforth, space by itself, and time by itself, have vanished into the merest shadows and only a kind of blend of the two exists in its own right." This is another example of writing upon which I cannot improve.

In physics, we usually start off with certain conventions, such as, which way is positive, where the origin is, and so on. As long as we do not switch conventions in the middle of a calculation we are free to choose these things how we want. An example of this is Minkowski space-time, which can also be defined according to $s^2 = -x^2 - y^2 - z^2 + (ct)^2$. It does matter which one you choose, as long as you don't switch horses in midstream.

I never understood why it is literally bad to switch horses in midstream, but then, growing up in New Jersey, I never rode a horse. The entire world came to understand the figurative meaning when it was discovered that the theoreticians used a computer code that contained one Minkowski space-time, and other parts of the calculations used the other form. After correcting the embarrassing error, the magnetic moment was recalculated, and soon came to be in good agreement with theory. Terry and I published our paper anyway, and he received his PhD for his contributions. Someday when the measurements get better, our work may prove to be more important, but for now the Standard Model is still in business. Without torsion.

Another fascinating problem that persisted for decades was the solar neutrino problem. If I were writing this book in 1990, this would be a main chapter. To get to it, we have to travel deep into Homestake, an abandoned gold mine in Lead, South Dakota. By 1970, physicists, under the direction of Raymond Davis, carefully hauled 615 tons of cleaning fluid a mile below the surface. Cleaning fluid is perchloroethylene ($C_2Cl_4$), the important ingredient being chlorine (Cl). The reason is this: When a neutrino slams into chlorine it can change a neutron into a proton plus an electron (this is another example of the weak nuclear reaction). Now, chlorine has 17 protons in the nucleus, and the next element on the periodic table is argon with 18 protons. Therefore, a neutrino can change a chlorine atom into an atom of the inert gas argon. By counting the number of argon atoms, we can count how many neutrinos passed through.

The 18 protons in argon like to be accompanied by 22 neutrons, giving it 40 nucleons altogether, but the newborns I just described have only 37—three neutrons shy. This is a radioactive form of argon, but it is detectable, and for 25 years physicists measured the amount of argon that was created. By the way, the reason for the extreme depth was to insulate the tanks from the ubiquitous cosmic rays, which would mess things up more than an entire school of first-graders.

The reason so much chlorine was used is due to the extreme weakness of the interaction. In other words, the probability that a neutrino will instigate this reaction is rare. But from where are these neutrinos coming? The sun. For every helium atom formed, two neutrinos are emitted. Because we know how much energy the sun gives, we know how many helium atoms are formed per second, and thus how many neutrinos are emitted per second.

In order to discuss these experiments, the solar neutrino unit (SNU) was invented: One SNU is one reaction per second for every $10^{36}$ target atoms. The 615 tons of perchloroethylene contains about $10^{30}$ chlorine atoms, which means we get about three reactions per month! Actually, Homestake was reporting an event rate at about 2.5 SNU.

The problem with this is that the theory predicted a value twice as high as this, and the three decades following 1970 had all of us up nights, wondering what in the world was wrong. The long and detailed calculations that describe fusion in the center of sun were checked and refined, the Standard Model was put back under the microscope, and quite a number of ideas were floated on the agitated sea of discrepancy. Other solar neutrino experiments were set up, including Kamiokande in Japan and the SuperKamiokande, all predicting that the observed flux was too low, and in fact too low by a factor of nearly three. Then in 1998 the interpretation of the SuperKamiokande data gave evidence that the neutrino had a mass of 0.1 eV, or more. This was spectacular, because the neutrino mass was originally assumed to be zero, like the photon.

I mentioned that there are three kinds of neutrinos: those associated with the electron, the muon, and the tau. Assuming these are massless neutrinos, they are forever locked into their families as rigidly as the Capulets and Montagues. However, if they have mass, no matter how small (but not zero), they are allowed to switch families as easily as movie stars. We call this *neutrino oscillations*. In the reaction I described earlier, in which a neutrino slams into chlorine to make argon, these are electron neutrinos, the kind made in the sun. If, on route from the center of the sun and all the way to Earth, the electron neutrino can jump families, turning into a muon neutrino, it will pass right by the chlorine atoms as if they were ghosts. This also explains the factor of three, if the electron neutrino has equal probability of switching into the other two kinds. This is now the commonly accepted solution to the solar neutrino problem, and the Standard Model, which originally had massless neutrinos, was patched so that it can accommodate massive neutrinos.

It is remarkable to me to see such a vast discrepancy in masses of the elementary particles. The electron is roughly a thousand times the mass of the neutrino, the mass of the up quark is a thousand or so times greater that the electron, and the charmed quark has a mass much bigger than that. Although some ratios of masses of particular particles can be predicted, the Standard Model does not predict the "mass spectrum" of the elementary particles, so it definitely has its limitations. Perhaps the biggest problem, of course, is that gravity has been ostracized from the entire family, and it seems that no amount of patchwork will allow it to be adopted.

Despite warm epochs in which confidence began to melt, the Standard Model, like the old clunker, is still on the road. The most definitive piece of evidence, however, has been the most elusive. The Higgs particle has not been detected. How exactly can you find the Higgs? First of all, particles are not like Captain Kidd's treasure, buried deep in some arcane location, nine paces south of the gnarled oak, and again three paces east. We are not hunters; we are parents. We do not look for it; we create it. So, you ask, how do you give birth to such a thing as a Higgs particle? The Standard Model gives the recipe: If a proton and an antiproton are slammed into each other, the Higgs particle will be created.

Many measurements depend on the results of many other measurements, and this is especially true in the Standard Model. Nowadays the best estimate of the mass of the top quark is 171 GeV. The best value for the mass of the W boson is 80 GeV. When these number are used, the highest value of the mass of the Higgs is 144 GeV. So we are closing in: Experiments indicate the mass of the Higgs particle is more than 115 GeV, and theory now tells us it is less than 144 GeV. *With better instruments come better observations, and with better observations we begin to understand the enigma that is our universe.* A workhouse in this quest to find the elusive Higgs particle is the Tevatron collider at the FremiLab in Batavia, Illinois. The proton-antiproton slam-fests at the Tevatron have a total energy of 2 TeV (2,000 GeV), and many believe it is only a matter of time before evidence of the Higgs particle begins to pile up.

What would happen if the Higgs particle is never found? I do not know, but, as the Canadians say, be prepared, not scared. Other theories have been looked into, including unified theories that started with five fundamental particles: the SU(5) theories. In this theory the free proton will decay (an event never observed). Another approach consists of assuming that the quarks are made from even more fundamental particles, called *preons*. There is no experimental evidence, and, as far as I know, no theoretical evidence for preons. Pretty soon you get the feeling there is no end to this process, but if there is some absolutely fundamental particle from which all others are made, I have the name ready and waiting: the *on*.

In summary, creating and detecting the Higgs particle is probably the biggest issue in particle physics. I can certainly recommend Leon Lederman's book that is dedicated to this particle (see the Suggested Reading list ). Of course, maybe the whole idea is wrong. Years ago people believed in a substance called *caloric*. It was somewhat similar to how we think of electricity nowadays, a substance that flowed from hot bodies to cold, warming them up. If you touched a hot coal, the caloric would flow from the coal into your finger. Now that your finger had additional caloric, it felt warmer. This theory had predictive value too, and was believed by many physicists for many years, but now we smile at it as ancient lore, because we know there is no such substance as caloric. As I explained in Chapter 2, the only difference between hot tea and cold tea is the speed of the molecules. We may even think of heat as an "emergent phenomenon," a property that results from characteristic properties of something more fundamental—in this case, the molecules. Perhaps someday soon we will have a comparable revolution in particle physics, and future physicists will smile on the Standard Model as ancient lore.

Or perhaps someday soon we will find the Higgs particle.

## Technical Note

If seeing a lot of number makes you faint, you better skip this note. Otherwise, let me talk a little about 2-by-2 matrices. Such a matrix is a collection of numbers, such as

$$A = \begin{pmatrix} 1 & 2 \\ 3 & 4 \end{pmatrix} \text{ and } B = \begin{pmatrix} 4 & 3 \\ 2 & 1 \end{pmatrix}.$$

The elements are simple numbers we can denote as follows
$A[1,1] = 1, A[1,2] = 2, A[2,1] = 3, \text{ and } A[2,2] = 4,$
and similarly for $B$.

We define matrix multiplication by $C = A \times B$, where:

$C[1,1] = A[1,1] \times B[1,1] + A[1,2] \times B[2,1],$

$C[1,2] = A[1,1] \times B[1,2] + A[1,2] \times B[2,2],$

$C[2,1] = A[2,1] \times B[1,1] + A[2,2] \times B[2,1],$ and

$C[2,2] = A[2,1] \times B[1,2] + A[2,2] \times B[2,2].$

Using these rules you can show that

$$C = \begin{pmatrix} 8 & 5 \\ 20 & 13 \end{pmatrix}.$$

I know I have gone into a lot of detail, but one of reasons is to show you that matrix multiplication does not commute, $A \times B$ does not equal $B \times A$. To prove this, I will let you try to show that

$$B \times A = \begin{pmatrix} 13 & 20 \\ 5 & 8 \end{pmatrix}.$$

## – *Chapter Six* –
# Quantum Gravity

## The Classical Theory

The first summer after I started graduate school my advisor told me to study for the qualifier exams. These are written exams, scheduled for four days, eight hours each day. They cover electricity and magnetism, quantum mechanics, thermodynamics (statistical mechanics), and classical mechanics—everything a growing physicist needs to know. I agreed, but told him I wanted to quantize the gravitational field first. He started banging his pipe, but I was undeterred.

I spent the entire summer working on it, and was very excited when September brought New York State's first frost. I went into my advisor's office and filled one board after another until the banging was so loud I had to stop. My advisor told me to look up some old paper by Rosenfeld, whoever he was.

I went to dust-laden shelves where journals of the 1930s sat unmolested for epochs. Like an archeologist, I dug through a foot of sediment to unearth the ancient writings. They were surprised to see me, but happy to show me Rosenfeld's paper—quantum gravity all worked out. Part of me was excited, to see that I had done a lot of things right, but a bigger part deflated, because it was already done (and I still had to study for the Damoclesian qualifiers). More important than any of this, I soon realized, was that Rosenfeld's paper was only an approximation. It was the first significant effort, but it was only meant to be a baby step. The full theory had major unsolved issues. Looking back now, this paper serves to indicate how extraordinarily difficult this issue is.

Let me tell you a little about gravity, the theory of general relativity, before we try to quantize it. Einstein saw gravitation as geometry. As opposed to electromagnetism, in which a charge creates a field and lines of force, matter curves space. Instead of using Newton's law of motion, Einstein postulated that a particle in a gravitational field will travel along a geodesic—the shortest distance between two points.

It is a blistering hot day in summer. I am cutting the grass, wondering if anybody else has noticed that the sun's energy output has tripled, as I spy some new structure in the backyard. My kids call it a trampoline, but I call it a sure way to break a neck. The good news is that I have less area to mow, but now that I am cooled off I can use it in a better way, as an analogy. If you flick a marble athwart the taut nylon it will roll across, traveling in a straight line. But if you place a bowling ball in the middle and then roll the marble, it will be deflected. If the bowling ball is heavy enough, the marble may not even make it to the other side, being trapped in the well of the ball. In between these two extremes is an orbit, such that the marble rolls round and round the bowling ball.

This is how Einstein viewed the universe. The sun curves the space around it, and the planets are traveling in the shortest paths possible. Actually, Riemann had this idea years before, but he was thinking in three dimensions. Part of Einstein's genius was using the four-dimensional space-time of Minkowksi I discussed earlier. This is impossible to visualize, but in a four-dimensional space-time continuum, in which time is somewhat like a fourth dimension, all the gravitational motions we see—whether it is Mercury racing across the sky, or a bad manuscript you toss into the trash—are described by geodesics.

Other forces cannot be viewed this way, which is one harbinger of why it is so difficult to quantize gravity. Another problem is that gravity is nonlinear. As a counterexample, if you look at the equations that govern electromagnetism in a vacuum, the field is never multiplied by itself. Never. This is a linear theory. In the field equations for gravity, the field is multiplied by itself all the time, giving a nonlinear theory. What is the difference between a linear theory and a nonlinear theory? It is the difference between putting water or gasoline in your tank, the difference between Kool-Aid and tequila.

Another point I would like to make is this: Gravity is the strongest force. The reason I say this is because all the books say gravitation is the weakest force. So, why am I cruisin' for a bruisin'? If you look at the force between two electrons, the electric repulsion is 35 orders of magnitude greater than the gravitational attraction. That's 100,000,000,000,000,000,000,000,000,000,000,000 times greater. At a fundamental level, where we like to compare—and unify—the fundamental forces, gravity is, in fact, hopelessly overpowered. Actually, if I write a sequel, I should probably have a chapter called "Why is Gravity So Weak?" But, on the large scale, gravity wins out. Consider yourself as you sit on a chair. The mighty nuclear force, the strongest, folds like a bad hand of cards. Gravity pulls you down and electric repulsion holds you up. So you see right there the gravitational force equals the electric force, and all those zeros are for naught. The reason is that charge can be negative and positive, and for you and the chair, the net charge is about zero, but gravity always adds. In fact, if a gorilla sat in the chair it might break, showing that gravity is stronger than electromagnetism.

If you have enough mass, gravitation beats out everything else. This is not simply a jeremiad down sophistry lane—I am coming to a point. Gravity is attractive, pulling itself together. Did you ever ask yourself why the planets are spherical but asteroids are potato-shaped? Gravity. Planets are much bigger, and the gravitational field is much stronger, forcing them into spheres. If you put two or three times the mass of our sun together, which happens after a large star ceases undergoing fusion, it collapses, and nothing can stop it.

I told you earlier about the fate of our sun, which will become a big hunk of carbon, cool off, and fade from sight. The electric repulsion of the electrons in the carbon atoms balance the gravitational self-attraction, and this happy equilibrium will continue until the cows come home. But if the mass were a few times bigger, the gravitational force would overpower the electric force, and the object would collapse. And nothing can stop the collapse. It will get smaller, and smaller, and smaller, until it is the size of the period at the end of this sentence, and then it will get smaller still.

My first course in general relativity was given by Hans O'Hanian, one of John Wheeler's students. He wrote an excellent book, but one thing I will never forget is the quote he used to introduce the chapters on black holes: *Abandon ye all hope who enter here*. Years later I was staying with my friend Warren Rosen when I was out of house. I saw Dante Alighieri's *Divine Comedy*, the origin of this quote, on his bookshelves, and was spellbound, missing lunch and dinner to finish it, but I digress.

When I interviewed a black hole I was very nervous and had to cut the interview short.[1] This is because, surrounded by this "point" into which all matter falls, there is an event horizon, a surface of no return. Once you get within the event horizon, forget about it, you can never return. Dante did not know the theory of gravity, but he knew about crossing the point of no return.

I am trying to stress how different gravity is from other forces. Take this business about curving space. Because we are really talking about space-time, this means matter curves time. What does this mean? One example is that time slows down in a gravitational field. Atoms near the surface of the sun, being in a stronger gravitational field than those same atoms on Earth, emit a slightly longer wavelength because of this effect.

Let's look at a prediction of general relativity. Let $t_s$ be a short interval of time on the surface of the sun, and let $t$ be the interval far away. Remember the bowling ball on the trampoline: Near the bowling ball the (two-dimensional) space is curved; far away it is flat. The gravitational field of the sun slows time, but far away time is unaffected.

This relation is expressed by

$$t_s = \frac{t}{\sqrt{1 - \dfrac{2Gm}{c^2 r}}} \qquad \text{(Equation 5.1)}$$

where $m$ is the mass of the sun, $r$ is the distance from the center of the sun, $c$ is the speed of light, and $G$ is the gravitational constant. I give you some more details in the Technical Notes at the end of the chapter, but if you crunch the numbers, this shows that for every million seconds that go by on Earth, one million and two seconds go by on the sun. This means that, when atoms emit light, the period is just a little bit longer, which means that the wavelength is a little bit longer. This is called *gravitational redshift*, and has been measured.

In this example, the quantity $\frac{2Gm}{c^2 r}$ is about one over one million, or about $10^{-6}$. Imagine if the mass could be squeezed into a smaller region, so that $m$ remains the same but $r$ gets smaller. This ratio will become larger, and, if you can squeeze it down far enough, it can be become 1. This means that the denominator in the formula becomes 0, and because 1/0 goes to infinity, this means that $t_s$ becomes infinity. Physically, if you think of a limiting process as $\frac{2Gm}{c^2 r}$ approaches 1, it implies that, on the surface, the time goes ever slower until it finally stands still.

This is a mathematical way of describing the event horizon of a black hole. For the mass of the sun you would have to squeeze all the matter (which occupies a sphere of radius 1 million miles) into a sphere of radius 2 km. Although gravity is pulling itself together—self-squeezing, so to speak—the nuclear processes that give off an enormous amount of energy produces the outward pressure that holds gravity at bay. When the nuclear processes come to end (which they will), nothing can stop gravity from pulling everything together. For a star that starts out with a few solar masses, collapse is inevitable. For big stars, black holes are as inevitable as sunset on Earth.

I would like to mention one more fascinating prediction of general relativity. Because gravity curves space and time, anything that travels through space and time is subject to this curvature, even light. Einstein was able to deduce this even before he formulated his general theory of relativity! A longstanding issue had concerned whether light was affected by gravity. Newton thought of light as being corpuscles, and physicists were able to deduce from this that light could be deflected in a gravitational field.

An early proponent of general relativity was Sir Arthur Eddington, who knew that starlight just grazing the surface of the sun would be deflected. But how can you measure starlight just grazing the sun? The sun is so bright it is impossible to see the feeble rays of distant stars. People were just as smart back then as they are now, and a rash of expeditions were initiated that brought people and telescopes around the globe, searching for an eclipse.

By the way, all eclipses are not created equal, and scientists dug through the charts to find one at a time when a lot of bright stars would be very near the sun. They hit pay dirt in 1919, a famous year that all general relativists to this day hold dear. The shadowy finger of the moon cut its black swath from Africa to South America, and the 1919 eclipse had scientists ready and waiting. A wonderful account of these times is given by Eisenstadt.[2]

Eddington went to Africa and another team went to South America, but the data was far from perfect. They had to decide if there was no shift, the "Newtonian" value, or the Einstein prediction, which is twice the Newtonian value. Some scientists were actually rooting against Einstein, and, due to the variations in the data, the controversy continued for quite some time. This experiment has been repeated many times since, using light and radio waves, and nowadays there is no doubt: Einstein was correct.

This is considered to be a turning point for Einstein's theory, but I see it as a metaphor for Einstein's theory. The physicists and astronomers trekked deep into remote sections of the wilderness to make these measurements, and thus were isolated, just as for years many general relativists worked in isolation: Not funded by industry, and with only small pots from other sources, general relativity failed to become a mainstream theory for decades. The impossibility of quantizing the theory underscores this fact, maintaining its isolation.

When we quantized the electromagnetic field, we named the quanta *photons*; for gravity we have reserved the name *graviton*. But even this may be too simplified a picture, so let us take a look at some of the quantum issues.

## Quantum Gravity

Perhaps it is not so surprising that, when it comes to quantization, gravity is obstreperous. Let me list some fundamental properties of quantum mechanics without gravity weighing things down:

1. Quantum mechanics is characterized by known operators.

2. These operators operate on the Fock space to give observable quantities.

3. The operators do not always commute ($AB$ does not equal $BA$), but we know what the difference is.

This last point may be written as $AB - BA = f$. We know what $f$ is, and this relation is called *the commutation relation*, and lies at the heart of quantum mechanics. The physics underlying commutation relations is extremely important: It is telling us that the measurements $A$ and $B$ interfere with each other. If they were completely independent, and could not affect each other in any way, then they would commute.

When it comes to gravity we:

1. Do not know what the operators are.

2. Do not have a Fock space.

3. Do not have commutation relations.

You cannot play soccer without a ball, and you cannot do physics without the light cone. Let me explain. In spacetime we talk about events, which means the time and place something occurs. I could clap my hands at noon in New York City, and one second later Katie could tap her foot in London. In principle my clap could cause her tap, and we say these two events are causal, or causally related. On the other hand, if I clap my hands at noon in New York City, and one second later Jennifer taps her boot in the Sea of Tranquility on the moon, these are not causal. The rule is, if the separation in time between events is less than the amount of time it takes light to travel between them, they are causally related (it takes about a second and a half for light to travel from here to the moon). Every particle has a light cone, which is the space of all events that are causally related. In my examples, Katie is in my light cone, but Jennifer is not.

When we try to understand the universe in which we live, we do so using models and theories, which consist of formulas. We have to start somewhere, so we are forced to make a few fundamental assumptions. One of the most fundamental assumptions we make is causality, which states that if $A$ causes $B$, then $B$ must come later than $A$. If $A$ and $B$ are not causally related, they cannot affect each other, and therefore they must commute. Now we know something about $f$: it must vanish outside the light cone. This idea lies at the very foundation of quantum mechanics, and is a nice way of saying that if $A$ and $B$ are not causally related (in other words, are outside each other's light cone), they cannot affect each other.

My editor put a limit on the number of words in this book, so why am I using so many to explain that $f$ must vanish outside the light cone? Because, in general relativity, mass curves space-time, and the light cone is distorted. We need commutation relations to define the theory of quantum gravity, but we need gravity to determine the structure of space-time, and the light cone. This is worse than a catch-22, but to the credit and genius of many physicists, it does not stop progress.

One of the earliest ideas came from Hermann Weyl, who reasoned that space-time itself must be quantum in nature. Sometimes I am up half the night trying to imagine what this means. Think about the number line: 0, 1, 2, 3,.... We know that between any two numbers there is another number, and so on. But what if the number line were not continuous, and came in discrete steps, similar to the board in Chinese Checkers? Time would be quantized too, jumping from one instant to the next in discrete intervals. In between these instants is nothing we can imagine.

**Question:**

*How can anyone think space and time are quantized? No measurement has ever shown it to be quantum in nature.*

☐ 1. We have not done the right kinds of experiments.

☐ 2. There is very little funding for research in quantum gravity.

☐ 3. The quantum units of time and space are so small we are unable to detect the quantum nature of space-time.

Concerning 2, the Department of Energy funds some research here, and the National Science Foundation shells out a few bucks too, but this, in my book, is definitely part of the answer. We can all agree on 3 though. Let me explain.

Although we do not have a quantum theory of gravity, we can speculate on the length and timescales are that are involved. We do this by using dimensional analysis. For example, if space is quantized, then there is a "smallest" distance. Max Planck guessed at what this smallest length should be, and we call it $L_p$. We assume Planck's constant must be involved, and because it is gravity we expect the gravitational constant $G$ is in the mix, as well as the speed of light $c$. Now we can play a little game: How can you multiply and/or divide these numbers and come up with something that has dimensions of length? I give you the gory details in the Technical Notes, but the answer turns out to be about $10^{-33}$ cm, teeny weenie indeed. This is the smallest size. Nothing smaller makes any sense; nothing smaller exists.

That is one way to look at it, but, if you ask general relativists, we would remind you of another way. You can also make up a length if you play with the constants that appear in general relativity, letting Planck's constant sit on the bench. If you look at the constants that appear in the equation on page 142 you can see that $\frac{2Gm}{c^2}$ has dimensions of length, which may be called $r_h$, the event horizon. To get the smallest "length," suppose we put in the smallest mass: the electron mass (we would get something even smaller if we use the neutrino mass, but this argument was founded before we knew the neutrino has mass, so I will stick to the electron). This gives $r_h$ equal to about $10^{-55}$ cm, 20 orders of magnitude smaller than the teeny-weeny Planck length.

This idea is a small step forward, perhaps, but there is a long way to go to formulate a quantum theory of gravity. In the 1960s some of the greatest physicists we had attacked this problem. They included Paul Dirac, John Wheeler, and many others, but they failed. They did make progress, and we now have a deeper understanding of both gravity and quantum mechanics, but no quantum theory of gravity. One formulation gave us an equation called the Wheeler DeWitt (after John Wheeler, and Bryce DeWitt) equation. It was a huge step, but to show you how hard it is to understand, it predicts that nothing moves. Now, when you get a theory that tells you nothing moves as you walk over to the coffee bar, you realize the theory needs more work. It came to be called the *frozen time formalism*, and, although we were able to reinterpret this equation to make it more sensible, it did not pan out to be the grail for which we search.

Earlier I wrote that the difference between a linear theory and a nonlinear theory is *the difference between putting water in your tank versus gasoline*. Let me explain further. In electrodynamics (linear), charge creates the field. In the quantized version, charge creates photons, but photons do not have charge. In gravity, mass creates the field. But because mass and energy are equivalent to the extent $E = mc^2$, and because the field contains energy, the gravitational field creates yet an additional gravitational field, so to speak. In a quantized version, mass creates gravitons, but gravitons create gravitons, and those gravitons create additional gravitons, and so on.

If we try to set up a quantum theory using the Fock space with bugs, you end up with an infestation that the Ghost Busters could not control.

Because Einstein viewed gravity in geometrical terms, let me review geometry. Euclid gave us the longest-running construction, a rigid structure with perpendicular and parallel lines we all learn about when we are too young to care. In 1908 Minkowski gave us a four-dimensional space-time, but it is a geometry, like Euclid's, that stands aloof, unbending and unyielding to everything except our imaginations. All of quantum field theory is based in Minkowski space-time, and the thousands and millions of measurements we make in accelerators, confirming our most astute predictions, result from theories that dwell in Minkowski space-time. The Standard Model lives in the house Minkowski built. However, with general relativity we learn of a new geometry, Riemannian geometry, that twists and bends due to the presence of matter as readily as a suspension bridge made of twine, and the entire structure of quantum field theory collapses like the Tacoma Narrows.

Besides the conceptual problems associated with quantum gravity, there are also more tangible thorns that rip through the scrim we call a theory. One is renormalization. I described that infinities appear in quantum mechanics, and we learned to avoid their sting by carefully balancing them against other infinities. But this approach fails doubly in quantum gravity. One reason is that the existence of an infinite energy in general relativity implies that space-time is warped into a point, what we call a singularity. This does not describe the universe we see. Worse, even if we could juggle infinities, there would have to be an infinite number of them. What this really means is that each new order of perturbation would require another set of infinite quantities—the theory is not renormalizable; the theory is no good.

Let's look at this situation in the mirror, for a moment. Maybe the problem is not with gravity; maybe the problem is with quantum mechanics. Sure, quantum mechanics works, but there are at least two problems floating on the surface like red bobbers. First, we must resort to renormalization. Although throwing around infinities works in Minkowski space-time, it is much more problematic in curved space, so maybe what we really need is theory free of infinities right from the outset. The second problem is: It does not work. We cannot quantize the gravitational field (so far, that is, with luck some of the methods I describe following might prove this statement wrong). There are, I am happy to report, brave souls working on the formulations of quantum mechanics. There are, I am sad to report, not many of them.

A frisson of hope pulsed through the physics community in the 1970s with the fixed-background approach. In this approach we visualize space-time as "broken" into two parts, the (fixed) Minkowski space-time plus the spongy curved space-time. When we quantize gravity, we quantize the spongy part, leaving the fixed Minkowski part alone. This is mathematically sound and intuitively appealing, but many argue it is homemade, quantizing only part of space and time. There were considerable technical difficulties as well, and this approach lost some of its momentum as the next decade rolled in.

Then, in the 1970s to 1980s another approach swept across the physics community, something that, to me, had extremely appealing traits, but also some rather untoward aspects. I told you about symmetries, such as rotating a book, or the SU(2). These operations form what mathematicians call a *group*. The trouble is, there are two different kinds of groups: the groups that transform particles are one kind (compact), and the groups in general relativity are different (non-compact). In order to make a unified theory of gravity with the other forces of nature, people tried to combine the two different groups, but failed. Even worse (or maybe better?!), it was shown to be impossible to combine them, in what came to be called the "no go" theorem.

My old friend Frank, who was studying astrophysics, liked to say, "Nothing is impossible," just to trigger a debate on what that means. He was right about the "no go" theorem, but we had to cook up another symmetry, which is called supersymmetry. Unlike the Standard Model, supersymmetry is much more egalitarian. It treats the fermions and bosons as equals, and the symmetry operators blithely change fermions to bosons (spin 1/2 particles to spin 0 particles), and vice versa. This is heretical, but it cracks an assumption in the "no go" theorem, allowing a unification of gravity with the other forces.

According to supersymmetry, every fermion has a boson "superpartner," and vice versa. For example, the mythical graviton (spin 2 massless boson, quantum of the gravitational field) will have a spin 3/2 gravitino, and some particles, such as the neutralino, will consist of combinations of superparticles. The drawback of supersymmetry is that none of these particles has been found, which brings me back to the old argument: Does this prove that, right away, the idea is bogus and unphysical, or should we be more open-minded, allowing for the possibility that there is more to this universe than meets the telescope?

If they must exist (according to supersymmetry) but we do not see them, then how can this be? We may invoke the same idea that we used to explain why we have never seen the Higgs particles: They are too heavy. String theory does not set upper limits to masses, so they may be extremely difficult, if not impossible, to measure. A bigger restriction of supersymmetry as I have presented it so far is that it is a global symmetry, meaning it acted instantaneously across all of space, a definite no-no according to special relativity. But now we know what to do: follow the path Yang and Mills carved out, and make it local, creating new fields—the gauge fields. This was a very exciting development. So exciting, in fact, that it made you forget about all those mythical particles. The gauge fields turn out to be gravity! This is called *supergravity*, and it gets better. One of the early and well-known show-stoppers of gravitation is that it is non-renormalizable. In supergravity, there is an infinity from both the boson and fermion sector, but they cancel. This is sweeter than a cherry lollipop.

Unfortunately, as people grow older, many lose their taste for lollipops. Supergravity did not lead to a quantum theory of gravity, and some people began to sour on the exciting developments of such a promising start. Sometimes, in physics, we settle for being part right. Supersymmetry could be an example: It seems to do remarkable things, yet perhaps it is not enough. Perhaps we need to invoke supersymmetry in some other context. This is exactly what happens in string theory, which I discuss in Chapter 8.

Things grew quiet until the great hubbub of 1986 erupted with the pioneering work of Abhay Ashtekar, but before I mention that, I have to emphasize that there are many people who contributed, many of whom are great physicists, but I do not include them. This is not intended to be a historical or complete work; I am trying to get to the physics we understand—and more, the physics we do not—and I leave out many, many brilliant people in the process. For example, it is hard to mention Ashtekar without bringing in Carlo Rovelli, and then Lee Smolin, and now I am really stuck, so I will simply quit.

Ashtekar's great idea was in the operators he used. I wrote that we do not know what the operators are. Until Ashtekar, most people used operators they obtained using the same line of reasoning we use to get the operators of quantum electrodynamics. Ashtekar eschewed this approach and came up with an exciting new set of operators, and, from these, so-called loop variables evolved. This work is still in progress, and, although some of the initial clamor has faded, some important problems have been solved, and this program might have a lot more to tell us.

For example, although there is no Fock space, there is a space on which operators can act. This may not sound like much to you, but if you were marooned on a desert island, imagine how you would feel if you found a bottle of water. Another result (one of the most compelling results) is that operators give a discrete spectrum of areas and volume. From these we can build a foam-like structure of space-time, like the head on a Guinness. It is a little strange that we do not have operators that give us a discrete spectrum of lengths, but we take what we can get.

Another idea is that perhaps we should not treat gravity as an isolated field. To feel the impact of what I am driving at, let us slip back to the 19th century for a moment, when we had two important fields: the electric and the magnetic. Toward the end of the century we had a small collection of equations describing all of electricity and magnetism, but there was a problem, an inconsistency. James Clark Maxwell corrected one of the equations by adding the now-famous *displacement current* term. When he analyzed the set of equations with this correction, he predicted the existence of electromagnetic waves, and more, predicted that they propagate at the speed of light! This work showed us that the fields are not two independent quantities, but are really different manifestations of the same entity, the electromagnetic field. Without this unified view, we could not derive a sensible theory of electromagnetism.

This was not lost on Glashow, Salam, and Weinberg. It was finally realized that a stand-alone theory of the weak interactions was not happening, but through a unified theory of weak *and* electromagnetic interactions, the correct theory emerged. This might be the solution to quantum gravity. Maybe we have not been listening to what the decades of failure were telling us; maybe they are the chorus saying we must first unify gravity with the other forces. But there was a problem, and a big one. Let me explain.

The moral of the displacement current was not lost on Einstein either, and, long before we had quantum mechanics, Einstein grabbed his hiking boots and compass, and began a long and torturous quest for a unified theory. This was long before Glashow, Salam, and Weinberg, and long before Yang and Mills, so Einstein's approach was nothing like the modern ones.

His was a geometric approach. If you move into a new house but only use the kitchen, you are not getting your money's worth. Einstein's house was the geometry of curved space, and his kitchen was the Riemannian geometry. By making several restricting assumptions, Einstein cloistered himself in the scullery; he did not take advantage of the full value of the geometry. Another way of looking at it is that in Einstein's original 1915 theory of general relativity, using Riemannian geometry, there are a set of 10 functions we have to find. In non-Riemannian geometry, there are 104.[3]

Einstein knew all this, and part of his genius was in being able to live in the kitchen, to home-in on the core of the theory, making it simple enough to understand, solve its equations, and test its predictions. But soon after 1915, and for the rest of his life, Einstein roamed through the entire house, looking for a unified theory of gravity and electromagnetism within the framework of non-Riemannian geometry.

Einstein was not only knight on this quest. Irwin Shrödinger, a founder of quantum mechanics, derived a unified theory of gravitation and electromagnetism, but it was flawed: The photon had a mass. I think it is always nice when new theories predict new things, but they should be right. Of course, it might be true, reasoned Shrödinger, that the mass of the photon is not 0. Very close to 0, yes, but not quite. Based on the known magnetic field of the Earth, Shrödinger calculated the upper limit of the mass of the photon, a result that stood for quite a while.

Nowadays we think mass should be generated by the Higgs mechanism, which is what the Standard Model does, and so we assume that the photon mass must be 0. Einstein worked year after year, and although he made some important contributions regarding the equation of motion, the great unification he sought proved as elusive as the Holy Grail. Einstein was never comfortable with the implications of quantum mechanics, and some feel the times swept by him, making his entire approach old-fashioned.

Einstein died in 1955, and most of the little momentum in this approach went with him. He never saw the Glashow-Salam-Weinberg model, which gives us the contemporary view to unification, making Einstein's later work almost quaint. You see where I am going. At some future date, will our Standard Model, and the local-gauge theory approach, become a quaint approach of the old ways? I think so, but, in its dark hands, time holds the final answer.

So, let's not skulk in the doldrums of nostalgia; let us modernize. The idea is to formulate gravity as a gauge theory, and then use the success of the Standard Model to quantize gravity. This was what Utiyama was trying to do, and then Kibble did a little better, until Frederick Hehl took the lead in the gauge theory of gravity.

Quantum geometry, or, in general, quantum gravity, is not only difficult, but is impossible (some say), so that the only choice—even if the hardest—is to start anew. The problem becomes, where to start? Is there a theory of gravity that is quantizable? Is it even possible?

Have we tried everything sensible? Might a crazy and un-precedented approach yield nice results? This is told another day. And so, we hoped that a gauge theory formulation of gravity would solve the problem, but it did not. In a yard where grass is the good news and weeds show the failures, it is time to re-sod.

Through all the difficulties and dead ends quantum gravity brings us, it is difficult to avoid reverting back to the notion of quantum geometry. Opposite to Minkowski space-time, there is the view that space-time itself is quantum in nature. Let us review Euclidean geometry, Riemannian geometry, and quantum geometry one more time. Because we cannot envision four dimensions, let us consider a two-dimensional surface.

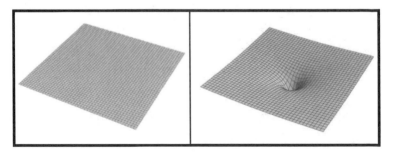

*Figure 6.1. Flat two-dimensional space.*

*Figure 6.2. Curved two-dimensional space.*

This geometry is supremely indifferent to its inhabitants. Bring in something as massive as a star and the geometry will be as unmoved as Ebenezer Scrooge. Take a microscope and examine this on a subatomic scale, or a subnuclear scale, and you will see the same smooth surface. This geometry was born carrying full inoculations to every germ of variation. A more accommodating geometry is Riemannian, an example of which is shown in Figure 6.2. Like the surface of a trampoline, this geometry bends and twists to accommodate matter and energy. The precise re-lation between the geom-etry and the matter is given by the Einstein field equa-tions. Bring in more mass and watch the curvature change. However, take a microscope and look at the subatomic scale, and it will

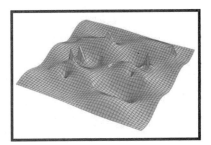

*Figure 6.3. A snapshot of space-time during its wild and restive dance.*

be smooth as a baby's cheek. Quantum geometry is expected to be different: It is a choppy foam, violent, and changing randomly with lightning speed. Figure 6.3 is an artist's (R.H. again) conception.

In truth, the space could be full of holes, and have handles that connect different regions with wormholes and knots.

Let me get your two cents on all of this with a question.

**?** **Question:**

*Which of the following is/are true about quantum gravity?*

☐ 1. Decades and decades of work by our greatest minds have failed; this shows gravity is not quantum in nature. All other fields are quantum, but gravity is different.

☐ 2. It is a waste of time. This issue does not keep me up all night, and, if settled, will not lower the price of gas.

☐ 3. We need to ask the aliens who are hidden away in New Mexico.

☐ 4. We need more funding to get more physicists to study this issue.

☐ 5. We need less funding; it is too hard anyway.

Number 1 comes up from time to time, but in our hearts most of us believe that answer is a weak cover for the white flag of surrender. However, it may not be surrender at all, and I will come back to this momentarily. I will not address number 2. Or 3. Or 5. But let me tell you a brief anecdote about number 4.

When I was a full-time university professor I would get a couple of graduate students each year telling me they wanted to study gravity. The first thing I would tell them is that they would have trouble getting a job after obtaining their PhDs. On their next visit, because I was chair of the department, I would tell them of other exciting work that was going on, and on their third visit I would repeat my no-jobs speech. Only if they came a fourth time I would consider working with them. This is the terrible state of affairs in physics: The single biggest unsolved problem is the worst area in which to major.

In order to understand another approach, let us briefly turn our attention to phonons. One of the many blessings of university life is the colloquium. Once a week esteemed physicists are invited to the university to give a talk about their work. We all do this, partly because it is an honor to be asked, and partly because it is part of the job. Living right in the middle of the University of North Carolina, Duke, and North Carolina State is like being in heaven, as far as colloquia go. Anyway, I was attending a talk on emergent phenomena and I felt as though someone just opened a window, giving me a whole new view of the world. Let me explain.

We know photons are the quanta of the electromagnetic field, but phonons? Remember that we live in a world that is quantum in nature, and, even though I limited my earlier writing to atoms or elementary particles, quantum mechanics also rules for solid materials, such as a copper wire, or all of those semiconductors we always hear about. With so many atoms involved, this problem becomes difficult indeed, and, as new materials egg on theorists, theorists actively work on better models to explain the behavior of these many body systems. In fact, some physicists may believe the subject of collective phenomena should be a chapter in this book, but I am wandering off point. The main result is that when quantum mechanics is applied to a bulk system, it predicts collective states resulting in quantized behavior. These quantum states, which consist of motions of many particles, are called *phonons*. An oversimplified analogy would be this: When the electromagnetic field is quantized, you get photons, and when a sound wave in a solid is quantized, you get phonons. My point is: phonons are an example of emergent phenomena. It would make no sense to try look for subparticles of phonons, because phonons result from the lattice of all the atoms in the solid. It would make no sense to build a theory starting out with phonons. The correct approach is to apply quantum mechanics to the underlying atoms, and let phonons emerge. In the Higgs chapter I mentioned anyons; these are another example of emergent phenomena.

This is old hat, but now let me show you that window I was opening. Suppose phonons are not the only emergent phenomena. What if electrons, for example, are emergent phenomena? This would gives us a totally new way of looking at nature, perhaps making our current standard views as accurate as the caloric. As long as we are putting our imaginations to good use, let us push even further. What if gravity is an emergent phenomena? In this case it would make no sense to try to derive a theory of quantum gravity, just as it would make no sense to try to derive a theory of quantum phonons.

This would imply that there is something more fundamental. If this is true, to repeat the overused metaphor in the preface, we have barely scratched the surface. We have set sail on uncharted seas, on unknown planets. It would revolutionize how we view nature, but it may explain why quantum gravity is the hardest problem there is: It is a problem that cannot be solved. It also explains why answer number 1 in my last question might be correct after all. In the chapter on string theory I will show you another indication of emergent behavior.

As a final thought, I will remind you that even the great physicists can be wrong. Two of the greatest were Wolfgang Pauli and Werner Heisenberg, who, in 1929, laid out some of the basics of quantum field theory, and said, "One should mention that a quantization of the gravitational field, which appears to be necessary for physical reasons, may be carried out without any new difficulties by means of a formalism wholly analogous to that applied here."[4]

## Technical Notes

### Planck Length

The gravitational constant is

$G = 6.67 \times 10^{-11}$ m$^3$s$^{-2}$kg$^{-1}$;

$h = 6.19 \times 10^{-34}$ mkgT$^{-1}$; and

$c = 3 \times 10^8$ m/s.

We define the Planck length by $L_p = \sqrt{hG/c^3}$, which is about $10^{-35}$ meters. The Planck time is the Planck length divided by $c$, which comes out to be about $10^{-43}$ seconds. Another cool number is density, $c^5/hG$, which comes out to be the extraordinary value $10^{94}$ g/cm$^3$.

## Gravitational Time Dilation

From the equation

$$t_s = \frac{t}{\sqrt{1 - \frac{2Gm}{c^2 r}}}$$

putting in the values of $G = 6.67 \times 10^{-11}$ and $c = 3 \times 10^8$, the mass of the sun $m = 2 \times 10^{30}$ kg, and the radius of the sun $r = 6.96 \times 10^8$ m, we have:

$$t_s = \frac{t}{\sqrt{1 - \frac{2 \times 6.67 \times 10^{-11} \times 2 \times 10^{30}}{\left(3 \times 10^8\right)^2 \times 6.96 \times 10^8}}} \doteq 1.000002 \times t$$

If 1 million seconds go by on earth ($t = 1,000,000$), then 1 million and 2 seconds go by on the surface of the sun. The gravitational redshift on Earth is four orders of magnitude smaller, and is ignored here.

*– Chapter Seven –*

# Ashes to Ashes—Things We Know

When I was in graduate school I only took one course in astrophysics per se, but it was enough to put me in the school of awe for the rest of my life. To me, it seems that astrophysicists have to know absolutely everything, from thermodynamics to quantum mechanics, from general relativity to the Standard Model: It takes all this knowledge to explain the birth and death of a star, and then some.

I will give an account of what we know, and end with a few points that are still in debate, but mostly this chapter is included to help you understand some of the exotic species that inhabit our universe. We start when the universe was a billion years old or so, and consisted of vast clouds of hydrogen, with helium spread across the vastness like whitecaps on a windy day. If you are single and go out on a lot of dates, you know that some are exciting, and some are as dull as those garden shears of mine. The universe at this time was like the bad date—nothing was happening. Until...

Gravity reigned over the entire kingdom, ordering its loyal subjects (hydrogen and helium) to muster, and they obeyed. They pulled themselves together in vast collections of smaller clouds that would become galaxies or clusters of galaxies, and within these, even smaller pockets would collapse to become stars. As the hydrogen cloud gets smaller and smaller, the atoms continually speed up because of the increased gravitational forces. As they collapse further they bang into each other all the time, but the collapse continues. Let us start with how our sun formed.

**Experiment:** You must do this experiment before you read on, but you can do it in your chair. Place your open hands together palm to palm. Squeeze them together very hard and rub them back and forth a few times as fast as you can. Afterward, go back to your reading.

This experiment shows how friction generates heat. At the atomic level (at the surface of your skin), when you rub, you are simply creating many collisions between molecules. Conservation of energy shows that the work you did to rub your hands is transformed into heat energy.

When I taught astronomy I did every demonstration under the sun, and one of my favorites was "the paper in the glass tube." A scrap of paper is placed in a glass tube that is fitted with a piston. I would speechify about thermodynamics, make them do the experiment you just performed, and then, after dimming the lights, plunge the piston home, squeezing all of the air down to a cubic centimeter or so, where the paper is. The paper bursts into flames as the class explodes in applause—a demo that worked.

The same thing happens as hydrogen collapses, although, because we are talking about the mass of the sun, the force is much greater, and there are many more collisions. Thus, the temperature increases by an enormous amount. We know hot things glow, so as the cloud gets smaller it will begin to simmer red. As the collapse continues it gets hotter and hotter, becoming bright white. This continues until, at the center, the core reaches 15 million degrees, at which point, kablooey: fusion begins. (Chapter 3 contains more details about fusion.) Soon the universe was ablaze with countless stars, shining like diamonds thrown across an ebony slate.

Let us step back and try to remember the Olympic high-dive. The Olympian is propelled upward, arms and legs as straight as a laser beam. Held in gravity's hand, she gently arcs high over the water until she tucks in her knees and arms, becoming a human cannonball. At that instant, she rotates so quickly that I can only see a blur. This is a good example of conservation of angular momentum: As she becomes smaller she must spin faster to conserve angular momentum.

I demonstrated this in my astronomy class, but, because my aquatic prowess never developed much beyond the belly flop, I took a different approach. I would sit in a chair that rotates, holding my arms straight out, clutching a 10-pound barbell in each hand. I would ask a student to give me a slight rotation, which he did with glee and much too much zeal. Then I would snap my hands into my chest and spin like a top. (You can do this yourself, with or without weights, by the way.)

Allow me another little detour, into the world of baseball, and then we can get back to work on formation—or, as the astrophysicists say, *stellar evolution* (they are very anthropo-morphic about stars). Years ago many great pitchers would house a chaw of tobacco, making their cheeks look like the central bulge of a galaxy. Occasionally they would impart some of the effluent on the ball. Chewing tobacco was banned around 1920 on sanitary grounds, but it was easy to detect miscreants: The umpire would have to wipe the brown sludge from his face. My point is: when the ball is thrown, the fluid peals away from the spinning ball. You can do the same thing in your kitchen, by spinning a wet apple in the air. Now, back to business.

These great clouds of gas could easily have a tiny net rotation as they begin to collapse. As they get smaller they must spin faster, and smaller parcels will rip away as the ro-tation rate spirals upward. These are the planets forming, and that is why I said, much to Kepler's chagrin, that the number of planets is simply an accident of formation.

The biggest chunk is Jupiter. Similar to the sun, it is mostly hydrogen, but, unlike the sun, it is a planet. Why not a star? In order for fusion to occur, two protons must be squeezed together until they practically touch each other. Positive charges repel, and the repulsive force spikes as they get very close (the force is proportional to $1/r^2$, so as $r$ gets small, $1/r^2$ becomes huge). In fact, the force is enormous, and it takes an enormous amount of matter to produce speeds high enough to overcome this barrier. The sun has enough mass to do this; Jupiter does not. Sometimes we think of objects like Jupiter as failed stars, or duds.

In fact, as the cloud of gas collapses, it is nearly impossible for pieces of the cloud *not* to be thrown off. This is why people think there are many, many more planets than the ones we see. In fact, it is likely that much bigger parcels of gas strip away from the shrinking, twirling cloud. It is also likely that many such parcels could be big enough to ignite fusion and become a star. It is also likely that when a star forms, two stars form, wildly rotating around each other like dancers in *Grease*. Binary star systems are fairly common, and, if you want to see one, just look at the Big Dipper (Mizar, the middle star in the handle, is a binary, but you need binoculars to resolve it).

In fact, if the star is not a binary, it probably has planets.

And now for my last two detours. When I was a graduate student I used to practice the fast draw with a pellet gun. After the $CO_2$ cartridge emptied, I would take it out. It was ice cold. Physicists call this *Joule-Thomson cooling*, and Floridians call it a godsend. It is the basis of how air conditioners and refrigerators work: Expanding gases cool. More precisely, expanding gases will cool as long as heat is not being added. If heat is added to a gas, then something else happens, which takes us back to Albuquerque, where the sky is dotted with brilliant-colored orbs. The hot-air balloons have propane burners, heating the air, expanding the air, making it essentially lighter than air. The moral of this detour is that, if you heat air, it will expand.

Let's go back to our sun, when fusion started in the core. The energy produced from the fusion explodes outward, creating what we call *radiation pressure*. The radiation pressure balances the inward pull of gravity, and a happy equilibrium is established. We estimate that this has going on for 5 billion years for our sun, and will continue another 5, until the hydrogen core is gone and we have a dense inner ball of helium. Then what? The radiation pressure that was fighting against gravity is gone, and the core collapses. This makes it hotter, so the hydrogen in the outer shell burns (fuses) even faster, making it hotter, so it expands, becoming as much as 100 times bigger than the sun. Then it cools, from the expansion, making its color change from white to red. This is the red-giant phase. When our sun becomes a red giant, global warming will be so strong that the entire oil industry cannot argue it away: The Earth will be within the sun.

During this expansion the helium core did not fall asleep. Its contraction caused it to heat up (remember the demo that worked), until it reached about 100 million degrees. By the way, you see the pattern that will continue throughout stellar evolution: Shrinking things heat up and expanding things cool (which is why those detours were so important). At that extreme temperature, the helium begins to fuse into carbon (don't worry about the electrons; they can take care of themselves, for now), or, as the astrophysics say, it undergoes helium burning. When the core is carbon, and after the outer layers have rocketed off, the star hits the end of the road. It is a hot carbon orb that eventually cools and fades from sight. The Cat's Eye Nebula, showing the remnant hydrogen-burning shell being ejected from the white dwarf, can be seen at *http://hubblesite.org/ newscenter/archive/releases/2004/27/*.

Things are even more interesting for heavier stars. In this case, the carbon is squeezed together with more force, creating higher temperatures, and even the carbon begins to undergo fusion. And it does not stop there; carbon to neon to oxygen to silicon until finally the great mass begins making iron at the core faster than all the steel mills in Pennsylvania (or should I say China?). The iron core grows and grows, but will not undergo fusion, because, for anything heavier than iron, fusion would *take* energy, not release it. These processes speed up, by the way. For a heavy star, "in approximate numbers, a star of 20 solar masses burns hydrogen for 10 million years, helium for 1 million years, carbon for 1,000 years, oxygen for 1 year, and silicon for a week. Its iron core grows for less than a day."[1]

The Earth has an iron core, which is fine for us, but for the end of a star, it is not quite terra firma. Remember: As the fusion comes to an end, the star will shrink. To exacerbate matters, the core gets so hot that all those atoms the star just made begin to break apart, which further accelerates the collapse. And then it gets even more dramatic. The protons and electrons are squeezed so close together that they combine, forming a neutron and a neutrino. This is more effective than spraying a wildfire with jet fuel, and the collapse is again accelerated. If the iron mass is about one solar mass or so, it can squeeze itself to the point that it is solid neutrons, but because it has so much inward momentum it collapses even further—a cosmic version of a bouncing rubber ball. The rubber ball gives us a fine game, but the bouncing star gives us something much better: the most spectacular explosion the universe has to offer.

It is called a supernova, and without it you would not be here to read this. The phrase *truth is stranger than fiction* is easily proven with this turn of events. In the rebound stage, not only does the star blow off all the outer layers of materials, but also, energy is so abundant that fusion into heavier elements occurs. Gold, uranium, and the other "heavies" finally enter our world. These too are blown free of the solid neutron core, and ejected into the vast space that makes our universe.

Millions or billions of years go by, and there are still great clouds of hydrogen collapsing into newer galaxies and stars. This time they are laced with stardust, the ashes of the supernova. The great cloud spirals ever faster, forming a star and flinging out neo-planets like a pitcher's spitball, but now they have more than hydrogen and helium: They have all the elements. And as a billion or so years go by, and the methane atmosphere becomes nitrogen and oxygen, and molecules form amino acids that form proteins, and the great oceans—the comet graveyard—spawns life, great bridges emerge, connecting bustling cities, and the supernova remnants rise up again.

The fact that these remnants rise up to see their creator, and remnants rise up to understand their creator, is nearly incomprehensible. But that is exactly what we are doing right now. As we peer into the past, measuring and characterizing the stars left behind by distant supernova explosions, we are studying, and understanding, our origins.

The stars that are the vestige of these supernova explosions are called *neutron stars*. They have the mass of a sun but are only 10 km or so in radius. Packed tighter than commuters on the Long Island Expressway, these are solid neutrons, similar to a giant nucleus (except there are no protons). One cubic centimeter of it, the size of a small sugar cube, would have a mass of $10^{12}$ grams. If you brought home such an object, you would have trouble keeping it. At 1 million tons, the little cube would break through your floor, smashing its way to the center of the Earth.

To understand how these exotic objects were discovered, we must go back to 1967, when I was still in high school. While I was being dragged into New York City for the specious reason that "everybody" was meeting there, other people were being more productive. The Be-in in Central Park was a thing to forget, but Jocelyn Bell, a graduate student at

Cambridge, made a thing to remember: a discovery no one could explain. She found that every 1.34 seconds a short blast of radio radiation would reach the Earth. (I do not mean she heard the soothing sounds of sonorous sambas, I mean that the wavelength of the radiation was the same as what we use for radio.)

Nothing like this was ever detected before, and the precise regularity was astonishing. The UFO-nut fringe had a field day, and some still believe it was aliens trying to make contact, although I would think their Environmental Protection Agency would put the kibosh on such a project.

There are two kinds of periodic systems this universe likes: orbits and spinning objects. Orbits involve long distances and long times, but things can spin much faster. The Earth takes a day and the sun takes almost a month (yes, the sun spins on its axis). If we remember our Olympian, we realize that if the sun could shrink to a much smaller size it would rotate much faster. In fact, if it could be squeezed down to neutron star, its rotation would be on the order of 1 second.

Bell's advisor, Anthony Hewish, figured all this out—and then some. The rapidly spinning neutron star, he reasoned, would have a very strong magnetic field. The sun has a modest field, and so would other stars. As one collapses into a neutron star, the field, squeezed into ever-smaller regions, becomes extremely intense. The final ingredient in this recipe is that the neutron star spins on an axis that is not aligned with the magnetic axis. This is like a poorly thrown football that wobbles because the spin axis (the axis from tip to tip of the ball) does not align with the rotation axis (the wobble). The spin axis of the football, as it goes through the air, sketches out a cone. The same thing happens if you spin a top: As it slows down it begins to wobble. Anyway, energy is emitted along the axis of the neutron star. Because it wobbles, every so often it points directly at us, so we see it. The energy is emitted continuously from the axis, but only when it points at us do we detect it. Thus, the energy from the neutron stars seems to pulsate, and so Bell discovered the first pulsar, and her advisor won the Nobel prize for his explanation.

Now we are able to step up to 1992, the year Pope John Paul II finally lifted the edict of the Inquisition against Galileo. He even threw in an apology. It was one year after Mount Pinatubo erupted, when Peter Collins was on the lower slopes of Flagstaff Mountain in Arizona, and, as was his custom, was staring into a moonlit sky made hazy by the recent volcano. He saw a star that did not belong, and when he looked again the next morning, it was brighter. Collins was not the first person to discover a new star. The list goes back for centuries, observers finding stars flaring to life, often to disappear within days. Some even outdo Lazarus, and rise again, and again, and again.

When Collins observed his nova (*new star*) we understood what was going on. Consider a white dwarf in a close orbit with another star. The stars are so close that the super-dense dwarf is able to steal some hydrogen from the outer layers of the companion star. The hydrogen smashes into the rock-hard surface, heating up continually as more of nature's favorite gas builds up, until—once again—kablooey. The hydrogen ignites, and for a few hours it is like looking into the very center of a star. Compared to the sun, or most stars, we see the 6,000-degree surface as downright cold compared to the 15-million-degree center. For a few days, nature shows us the heart of the beast.

Not all of the new hydrogen burns off, by the way. Thus, with each nova the carbon star gains a little mass. At some point gravity is so strong it crushes the carbon and the star "goes supernova." This is different than the supernova I already explained, so we call the carbon crushers a type-I supernova, and the non-binary kind that start off in life simply too big, type-II supernova.

We have come a long way since Galileo defied the Church and turned his crude telescope toward the heavens. Nowadays we do not just use the visible spectrum, but search with detectors that can detect everything from radio waves to gamma rays. By 1980, astronomers were measuring short bursts of x-ray radiation analogous to pulsar observations. We now think that this is an example of a binary star in which one is a neutron star. As the hydrogen accretes on its surface, fusion ignites, but because the gravitational field at the surface of a neutron star is so much stronger than that of a white dwarf, the fusion occurs at a much faster rate and is accompanied by emissions that are in the x-ray spectrum.

Many times in science we are looking for one thing and find quite another. Hamlet was right on the money when he told his friend, "There are more things in heaven and earth, Horatio, than are dreamt of in your philosophy." This takes us to October 22, 1979, when the National Security Council sent a classified memorandum that eventually wound up on Jimmy Carter's desk, adding to the troubles of his presidency— a presidency that was encouraging nuclear non-proliferation. The subject was "South Atlantic Nuclear Event," and the first paragraph began, "The Intelligence Community has high confidence, after intense technical scrutiny of satellite data, that a low-yield atmospheric nuclear explosion occurred in the early morning hours of September 22 somewhere in an area comprising the southern portions of the Indian and Atlantic Oceans, the southern portion of Africa, and a portion of the Antarctic land mass."[2]

The "Vela Incident," as it was called, was triggered by the satellite Vela 6911, one of a fleet of satellites launched from 1963 to 1965 to monitor nuclear tests that would violate the Nuclear Test Ban Treaty. A scientific panel was convened and headed by Nobel Laureate Luis Alvarez, who concluded the event was anomalous, an error caused by aged and faulty bhangometers—the detectors onboard the Vela satellites. The controversy is still going on, by the way, and as fascinating as it is, the bhangometers raised an even more interesting controversy. As Horatio said, "O day and night, but this is wondrous strange."

The "watching" satellites had gamma-ray detectors, and, in a paper published in the *Astrophysical Journal* in 1973, a study of the detectors for the previous three years showed a series of gamma-ray-detection events.[3] The actual blast was of short duration, a few tenths of a second, and, because the satellites were spread in orbit around the Earth, they were able to roughly triangulate on the direction of the blast: The bursts seemed to come from outer space. Not the sun or the center of the galaxy, but from all over, and no one could explain their origin.

*With better instruments come better observations, and with better observations we begin to understand the enigma that is our universe.* Things improved in the 1990s after the launch of the Compton Gamma Ray Observatory, which carried BATSE, the Burst and Transient Source Explorer instrument. This is a sensitive gamma ray detector that showed clearly that the GRBs came from everywhere and were isotropic across the sky. This left us with two theories: One is that they came from some strange event occurring in the halo of our galaxy, and the other assumed that they are not from our galaxy, but in fact are spread throughout the cosmos.

The problem with the second argument is energy. Suppose you are camping in the woods, and, just before you jump in your tent or sleeping bag, you see a light. You do not hear anything, and wonder: Is that a flashlight at the next camp-ground, or a powerful flood-lamp a mile away? The point is: if these objects reside in remote regions of our universe, then the energy would have to be enormous—$10^{47}$ joules. This would be similar to converting most of the sun totally into energy in seconds, and nobody could figure out how nature could manu-facture such huge outpours.

I remember these times well. I did not have a dog in this fight, and read with great interest papers that took one side or the other. I was hoping that the solution would be some totally new phenomena that was happening locally, but in my heart felt that the only thing powerful enough must in-volve neutron stars and/or black holes, which boast the stron-gest gravitational fields in the cosmos.

I do not want to overuse the *with better instruments...* quote, but it strikes again in 1997, the year Tony Blair was appointed prime minister of the United Kingdom, and, with nearly 42 years under his belt, Strom Thurmond of South Carolina became the longest-serving member in the history of the United States Senate. It is also the year BeppoSAX, the Dutch-Italian satellite, detected GRB 970228. It was able to point its x-ray cameras to the emission, giving a location good enough for observatories to aim their optical and ra-dio telescopes within hours. I loved this period, when scien-tists across the globe acted with the speed and accuracy of an great army, but instead of blight they brought light.

Astronomers were able to calculate the redshift and found that the GRBs were located in galaxies in the most remote regions of the universe (remember Hubble: the larger the redshift, the farther away the object). They have been measured to have redshifts corresponding to distances between 6 and 12 billion light years. When you stop to think that the universe is around 14 billion years old, you realize we are looking at events that occurred deep in the past, when our universe was barely adolescent and still ignorant of the full pleasures of life. This solved the mystery of where the GRBs are, so it was then necessary to derive a model of how nature could manufacture such gargantuan outbursts.

No known process can emit this enormous amount of energy, but that enormous number was calculated assuming the energy of the burst radiated isotropically into space, the way the sun radiates its energy. But if the burst is focused like a laser beam, the energy would not have to be so extreme. The commonly accepted view for at least some GRBs is the Collapsar model. (After a few sets of tennis with my kids, this also models my behavior.) This is sort of a rapidly spinning type-II supernova/pulsar combo, in which the material outside the black hole collapses, and, due to the extreme spin, forms jets, which focus the energy in beams. A somewhat peskier issue concerns shorter-duration GRBs. These are thought to be mergers of two neutrons stars, but that is still more speculation than fact.

Neutron stars and GRBs may seem to be wild objects, but they are tame compared to what happens if the neutron star is too big for its britches. If it is more than about two solar masses, the gravitational field is so strong that it overpowers the neutrons, crushing them and causing further collapse. This time there is no stopping the collapse, and we fall into a region where our physics fails. Everything falls to a point, a region with no volume, but all of the mass of the neutron star (except for what blew away). We define *density* as mass divided by volume, so if you try and compute the density you divide by zero—a no-no in the world of physics and math. We call it a *singular point*, or a *singularity*. If you calculate the curvature at that point, it too is singular. This is unphysical; this makes no sense; this is not something we can work with or predict things about. That is why I said our physics fails.

Things are not really this black. These results I described are the prediction of general relativity—classical general relativity. By the time all that matter approaches the size of an atom, we know (believe) that a classical theory cannot hold. Therefore we know (believe) that this business about physics failing is hyperbolic. We must stop dreaming of classical results and compute what is happening using quantum gravity. Aye, there's the rub: We do not have a quantum theory of gravity, as I explained in Chapter 6.

Nevertheless, there are many regimes in which the classical theory is expected to hold true, or very near the truth, and it is interesting to look at more of its predictions. Mathematics is the bridge between our imagination and our physical world. Often that bridge takes us to new and unexpected shores, and we cannot determine if they are physically real, or simply the predictions of a great mathematical machine that puts logic above reality. For example, suppose I tell you there are some apples in a basket. You cannot see in, but I give you a clue that the square of the number equals 9. You say, "That's easy, there are three," but mathematics gives us the equation $n^2 = 9$, and proceeds to give us two answers, $+3$ and $-3$, each of which solves the equation. Thus, when we yield to mathematics, we must wonder if its answer is valid.

In this simple example it is easy: You throw away the "spurious" answer of $-3$ because it makes no sense. Often, however, things are much too complex to know *a priori* if a solution is real, or if it is the figment of the imagination of mathematics. A case in point concerns the single most important exact solution to Einstein's field equations, discovered by Karl Schwarzschild in 1916. Because we can view Einstein's theory in geometric terms, I will give you some solutions in terms of geometry.

First, let us revisit our earlier equation. We saw that if

$$\frac{2Gm}{c^2 r} = 1,$$

time seems to stand still, but worse, a singularity (infinity) occurs. This seems quite unphysical, and was a point of concern and debate for decades. Some argued that you should not worry about it because such objects, where all the matter is squeezed into a sphere of radius smaller than the event horizon, do not exist. Others felt that everything ended at the event horizon, and space and time did not exist inside. Yet Einstein stressed more than anyone that physics does not depend on the coordinate system you choose, and the coordinates should be chosen to simplify the physics. For example, if you want to study the electromagnetic energy density in your microwave oven, you would choose Cartesian coordinates, mutually perpendicular coordinates $x, y, x$ that line up with the walls of the oven. If you are studying waves on a drumhead, you would choose polar coordinates that have a radius and an angle. The coordinates I used in the equation are called Schwarzchild coordinates, and although there were only a few isolated attempts at looking at other coordinates, Schwarzchild coordinates were like a healthy king, and none usurped their throne—until 1960, that is, when Kruskal discovered another set of coordinates.

Let me give you a simple example of a case in which the space is fine but the coordinates get you in trouble. Think about the surface of a perfect sphere, and lay out longitude and latitude lines as we do for the Earth. You will find a North and South Pole, but these points are not special at all; they result from how you laid out the coordinates. If I tell you to go the North Pole and pick out the longitude 110 degrees, you cannot, because all of the longitude lines pass through that point, and, at that point, you cannot distinguish between them. (The same problem comes up when you comb your hair, of which I have vague memories.) But there is nothing special about the North Pole, the coordinates you chose make it stand out.

The singularity at $\frac{2Gm}{c^2 r} = 1$ is of this nature, and in Kruskal coordinates this point is nothing special. These coordinates were very helpful in understanding the space-time outside the event horizon, but when you plot them on paper, a funny thing happens: The page is half empty! Mathematically, we can extend the Kruskal coordinates to fill the page. Remember: One coordinate system is as good as another, so the Kruskal coordinates may be telling us that by using Schwarzchild coordinates, we are missing half of the space. For example, I showed you a curved surface in Figure 6.2 described by Schwarzchild geometry, but this was only using half the space. The surface is shown in the entire space in Figure 7.1.

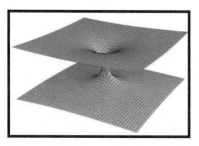

*Figure 7.1. Curved space with two flat regions far away, connected by the Einstein-Rosen bridge, or wormhole.*

This shows two regions that look like Euclidean space as you get far from the center. We call the regions asymptotically flat, but they are connected by a wormhole. The top region may be the space-time we observe, and the figure shows that there is another asymptotically flat region, the bottom region. This represents a full and complete space-time, but it is nothing we can see from this side. It can be another universe (whatever that means), or it may be a distant region of this universe, perhaps in another galaxy.

If we are on the top, then the bottom region is a white hole. It is called a *white hole* because, in principle, things can travel from there to the upper region. If there were a light there, the energy could flow into the top region, except for one problem: The wormhole pinches off too fast. This picture, however, is a "vacuum solution" to Einstein's equations, and there is fascinating speculation that, with the right kinds of additional fields, this wormhole could remain open.

I am not a big fan of modern science fiction (who needs the fiction?), but I have seen enough bad movies to know that sci-fi enthusiasts have glommed onto the wormhole concept. As fascinating as they are in that medium, they may be even more important on the subatomic scale where fluctuations in geometry rule the day. Wormholes may connect different points, making space multiply connected, and even change the topology of space. I imagine that many of our efforts, tried and true for Minkowski space-time, are effete against such a dynamic, twisted space as this.

In conclusion, I leave you with the question: Are wormholes real, or are they the figment of mathematics, as spurious as having a basket with −3 apples? The United States is riddled with research centers, often sewn together to form a fabric that weaves through universities, and sometimes stand alone. They cover everything from optics to low-temperature physics, and are funded by myriad sources, from the National Science Foundation to the Department of Defense. But there is no Center for the Study of Wormholes, and, unless we can drum up interest from the waste-disposal industry, there probably never will be.

## – *Chapter Eight* –
# String Theory

## Particles

I am standing on a low bluff, watching the Pacific Ocean pound a scraggly coastline festooned with Wax Myrtle and Coyote Brush, when I realize it is almost 9 o'clock. I scrabble across the windblown chaparral and get back to the Santa Barbara campus just in time to be only a little a late for the first talk. This is the Pacific Coast Gravity Meeting, back in the 1990s, when they alternated between Cal Tech and the University of California at Santa Barbara. They are very informal meetings, where graduate students are as plentiful as the sandals. When my talk is over, Ted Jacobson comes up to me and asks a question, so we go to a blackboard and fill it once or twice. Ted is telling me that my theory of gravity is equivalent to string theory.

Some people are excited about this, but not me. To show you how these things fit together I should back up a little and tell you that Einstein had to make a number of simplifying assumptions. This may sound strange if you think his theory is complicated but, believe me, it can get much more involved. Much of my research throughout the years has been looking into the generalizations of Einstein's theory. From a geometrical view, this kind of physics takes us from Euclidian geometry to Minkowskian, and then to Riemannian. Riemannian geometry allows for curved space, but has a couple of special assumptions that makes it simpler to deal with; when you relax these assumptions you have non-Riemannian geometry, as I explained earlier.

When you look at the physics of it, Riemannian geometry is characterized by one field: the gravitational field. Non-Riemannian geometry has two additional fields. My theory also has two additional fields, and they are, mathematically, the same as those that arise from string theory. This is what Ted was telling me. The trouble was, I did not believe string theory was a correct description of nature—and more importantly, I did not understand it.

The key to physics is simplicity and beauty, but those are relative terms. As an example, consider the elements, from hydrogen to uranium, say. An example of a theory that is neither simple nor beautiful is one that postulates the existence of 92 different kinds of matter. A simpler and more attractive theory assumes that atoms consist of only three particles: protons, neutrons, and electrons. The difference between the elements is simply the number and combination of these "elementary" particles that make up each atom.

Going from 92 to three is an example of simplicity, building up strikingly different kinds of elements, from the inert gas helium to the highly active metal lithium by adding one electron and one proton (and a couple of neutrons), is simplicity and beauty. Things did not stay simple for long, however, and complications arose in 1932. Let me explain.

I discussed Robert Millikan and his unusual form of torture, but I am sure you saw that I was not really being pejorative. He was a great physicist who applied his talent to everything from writing textbooks to antisubmarine warfare, and became the director of the Norman Bridge Laboratory of Physics at Cal Tech in 1921. He had a graduate student named Carl Anderson who graduated from Cal Tech in 1930 and ripped apart our simple view of the world two years later, winning a Nobel Prize for his discovery four years after that.

Let me digress with a point. Did you ever start driving early in the morning, and as soon as you get moving the windshield gets covered with condensation? You put on the wipers and swoosh it away, but within seconds it is back. This happens because the air is loaded with water. It is swamped with more water than it wants to hold, but has nothing handy on which to deposit it. The blades of grass do their part, dripping in silence until the sun bears down, warming the air until it can accommodate its precious gift, and now your windshield joins the crusade too.

Anderson was using a cloud chamber, which is just what it sounds like: a sealed apparatus with supersaturated water vapor, like the stuff you drove through. If a particle—say, a cosmic ray—zips through, it will leave a trail of ions, and each ion will act like your windshield, as a source upon which condensation can occur. The net result is a streaky line through the cloud chamber. Here is what Anderson saw.

Across the middle is a lead plate used to slow the particles down. The line curving down is an electron, but the line curving in the opposite direction must have the opposite charge. Carl Anderson published a paper in 1932 entitled "The Positive Electron," which opens with the following sentence: "On August 2, 1932, during the course of photographing cosmic-ray tracks produced in a vertical Wilson chamber (magnetic field of 15,000 gauss) designed in the summer of 1930 by Professor R.A. Millikan and the writer, the tracks shown in

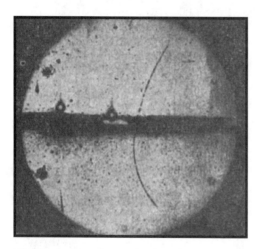

*Figure 8.1. Picture of cloud chamber with an electron and positron, from* http://en.wikipedia.org/wiki/Cloud_chamber.

Figure 8.1 were obtained, which seemed to be interpretable only on the basis of the existence in this case of a particle carrying a positive charge but having a mass of the same order of magnitude as that normally possessed by a free negative electron."[1]

This represents the first discovery of antimatter—the antielectron—which Anderson dubbed the *positron*. An electron and positron can be created at once, having equal mass, but opposite charge and opposite spin. The subsequent discovery of the antiproton and antineutron double the number of our modest stable to six, but this represents the door just after it was unlocked. Our solar system grew from the solid six planets that sufficed for millennia, and then our universe exploded into in unimaginable number of galaxies, and for the next few decades the world of elementary particles experienced a similar growth. A few years later Anderson discovered a negatively charged particle that has a radius of curvature somewhere between that of a proton and an electron. It seemed just like a heavy electron, and is named the *muon*.

I will not present a history of the discovery of elementary particles—that could be a book all by itself, and the meaning of *elementary particle* could be yet another book—but suffice to say that soon there became so many elementary particles, of such different character, that we began referring to *the zoo of particles*. With the need to deal with dozens of different particles, the beauty and simplicity of particle physics melted like ice cream at a summer picnic.

There is one point I can put off no longer. It was quite an issue, but it was solved so quickly it was like closing Pandora's box before anything flew out. In 1926, Irwin Schrödinger gave us an equation that serves as the basis for quantum mechanics. Some equations rock the boat, but this burst the entire ship asunder. When you solve it for hydrogen (nowadays a sophomore-level task for physics majors), you predict the famous emission lines I showed you in Chapter 1. The only problem was that 20 years earlier Einstein had published the special theory of relativity. That is a bad sentence. I should not imply Einstein's theory is problem; I should say the two theories were incompatible: Shrödinger's theory failed for particles that travel near the speed of light, the realm where special relativity is most important.

It was not that Shrödinger was unaware of special relativity; he was a smart guy, and in fact his first quantum mechanical equation was in perfect accord with special relativity. It had one problem, though: It gave the wrong answer, and physics can tolerate anything but the wrong answer. His second try is what came to be called Shrödinger's equation.

When I was an undergraduate I switched majors from engineering to physics, which was my version of the five-year program. The extra year gave me a lot of time, so I had many opportunities to wrestle with the Shrödinger equation. This is the most revolutionary equation in science, and predicts that everything we measure on the atomic scale is quantized, coming in discrete little bundles. It solved the mystery of the lines in hydrogen we saw in Chapter 1, and laid the foundation for our understanding of the behavior of semiconductors, leading to another revolution and the rise of the Silicon Valley. However, in this, his second attempt, Shrödinger eschewed the principles of relativity, and everyone knew the theory was not quite right.

Then comes Paul Dirac, one the most brilliant physicists of all time. He was the pastor who married quantum mechanics and special relativity, and he performed the ceremony in 1928. Even more amazing, after such a short courtship, the marriage has lasted all these years, giving birth to the theoretical basis of the spin of particles, something that was postulated earlier to describe the splitting of spectral lines. (I will talk about this in a moment.) But even this great union had issues, and one of them came up right away. It will take another paragraph or two to explain this, but it reveals an important attribute of great physicists (and possibly all great thinkers).

I told you that quantum mechanics predicts that energy is quantized. For example, when you apply the Shrödinger equation to hydrogen you get discrete energy levels. As the electrons hop from level to level they leave behind the line spectrum we saw in Chapter 1. Left undisturbed, atoms go into the lowest states, emitting photons as they do. This is why we see the hydrogen, as the current through the tube continually re-excites the atoms and just as quickly they jump back down.

Now, Dirac's theory gave quantized energy levels, but it also gave negative states, an infinity of negative states. This is clearly wrong; we do not see atoms falling into these negative energy states, and, as I said, physics can tolerate anything but the wrong answer. At this point, your average genius might throw in the towel, but not Dirac. He assumed that the negative energy states are already filled, and that we live in a sea of negative energy states filled with electrons. Now, if you believe this, you can believe that occasionally one of these electrons that abides in this negative sea can acquire enough energy to join the world we see, but this would leave behind a "hole," an empty state in the negative sea. But if you take away negative charge you make the "sea" more positive, and this "hole" is the site of the positive charge. In other respects the hole acts as an electron, except it is positive. Here is what Dirac wrote: "These holes will be things of positive energy and will therefore be in this respect like ordinary particles. Further, the motion of one of these holes in an electromagnetic field will be the same as that of the negative energy electron that would fill it, and thus will correspond to a charge + e. We are therefore led to the assumption that the holes in the distribution of negative energy states are the protons."[2] He was close, and is often credited with predicting the existence of the positron.

My real point is that great physicists believe in their work and push it to the limits, often making discoveries in the process. More importantly, I am trying to show you how physicists find simplicity and beauty. Instead of having to postulate two different particles, Dirac envisioned a simpler world, wherein the proton and electron were two different states of the same particle. This theme will come back to us many times. Later, the invention of quantum field theory obviated the need for this ominous infinite sea, but nobody can do everything.

Other kinds of theoretical predictions predicted predictable things, and experimentalists built accelerators big enough to enclose cities to create such particles, and this "zoo of elementary particles" (that phrase was popular years ago but has become unfashionable in recent years) lies fenced in only by the size of our machines and the extent of our imagination—but you see my point: The simplicity of a world with only three particles has, like the triceratops, become extinct. There is no use crying over spilled protons, so let us step back and try to find the true beauty and simplicity upon which we believe (hope) our world is based.

## Strings

It is easy to make two houses look different (just ask my neighbors). But it is harder to make two motes of dust look different. They are smaller and exhibit less structure. If you think of a particle as a point, with no length, no width, and no height, you can imagine how hard it is make such objects appear to be different from each other. For example, helium is different from lithium because of extra particles, but, if you have simply a point, you have no wiggle room, nothing to make one different from another. Therefore, if you have a quark in one hand and an electron in the other, and they are point-like particles, you are forced to postulate that you have two different particles, and not two versions of the same thing. Contrarily, all atoms are the same: collections of protons, electrons, and neutrons. You can say their difference arises from the number of particles, but you can also say their difference arises because they have different geometry.

If particles had structure, they could come in different shapes, and that, at least in theory, could describe how a single entity appears as two (or more) different particles. Imagine the simplicity of a world where there is only one entity, and all of the different particles we see are simply different shapes of that single thing. This is even simpler than the world of three particles—more beautiful too, in my book.

This is the string, a one-dimensional object that revolutionizes our world more than Starbucks. Because it is so revolutionary, yet so simple, let me emphasize it a bit. A string has one dimension: length, but no thickness. If you buy spaghetti as much as I do, you know there are the thick number-8s, the thinner number-9s, and the wispy number-10s, not to mention the ethereal angel hair. If you imagine number-1,000s, then 100,000s, and so on, you begin to see the string—one dimension only. Also, it is better to think in terms of cooked spaghetti, which can take on any shape. Here is an artist's conception of strings (and I am taking poetic license in using the word *artist*).

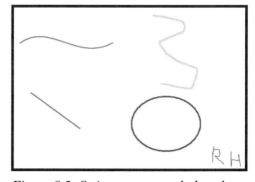

*Figure 8.2. Strings, open and closed.*

One of these may be an electron, another a quark, and so on. There is almost a classical quality to this picture. For example, you can play an A on a guitar by plucking the open string near the middle, and then again by plucking it near the bridge. Same note, same pitch, yet you can easily discern the difference (in the music a little *m* appears over the notes you are to pluck at the bridge, which stands for *metallic*, which describes the sound). What exactly is that difference? Why does one sound different from the other? In each case the string oscillates back and forth at 110 Hz (cycles per second), and if you are careful, they can emit the same total energy. The difference lies in geometry: The shapes of the vibrating strings are different.

Different shapes give different sounds; different shapes give different particles. Visualize an electron as a string that contains one wave. Now suppose it slams into something that has enough energy to change its shape, so that it contains two waves. This is relativity simple, if you ask me. Now translate this sentence to (point) particle language and you have the electron changing into another particle, something much more mysterious.

There is more good news. What happens if a string breaks? Say, in the figure, the top-right strings breaks, leaving the pieces on the left. What is a more natural way of describing how a particle decays, transforming into other particles? This is the simplicity and beauty I am talking about, and the good news does not end here. Earlier I described the benign tumor of renormalization. The infinites arise because you have to cope with something of zero size. The non-zero size of strings avoids these infinities, and the theory, it is believed, is free of this pathology.

Perhaps the biggest news came in 1984, so important it came to be called *the string theory revolution*. It was shown that string theory contains both gauge theories of the type found in the Standard Model *and* the general theory of relativity. I think this is what made people stand up and take notice. I told you earlier that quantum gravity does not exist, but if gravity falls out of string theory, a quantum theory with no infinities, the hardest problem of all time has been solved.

This is the emergent phenomena I was talking about, in which the fundamental thing is the string. Perform the math, set up commutation relations, and, like Henry Hudson, see where it takes you. In order for the theory to make sense there must be three fields: a scalar field, another that we call the antisymmetric field (or torsion), and a field characterized by spin-2 exchange particles—the gravitons. Starting from the basic idea of quantum mechanics and a string, gravity emerges. Very nice, in my book.

The good news is rolling in like waves on the Jersey shore, but a few crabs are in the mix, biting into the joy that is strings. Before I go into more details we must take another quick swim in the ocean of elementary particles. There are two kinds of people: women and men. You might enumerate other differences such as age or salary, but this one is quite fundamental. And so, there are two kinds of particles: fermions and bosons. Let me explain.

As physicists of the early 20th century looked more closely at the spectral lines, they saw that many were split apart into two or more lines. Sam Goudsmit, a physicist born in The Hague, was quite interested in this issue, and was able to derive little formulas that fit the data. Then he hooked up with George Uhlenbeck, who was also Dutch, and they realized that, if the electron had spin, all these empirical formulas could be put on a firm physical foundation. They sent a paper to their mentor, Ehrenfest, that described a spinning electron, and they sent off a paper to the journal *Naturwissenschaften*. But then they asked Lorentz , a preeminent physicist of the time, to have a gander at their paper. He then showed that such a thing is impossible: The electron cannot spin fast enough. Let me explain.

If you charge up, say, a potato, and then spin it, it will produce a magnetic field. In fact, the magnetic field is proportional to the spin. Uhlenbeck and Goudsmit figured this would hold for the electron as well, but the calculations of Lorentz brought down the kibosh like lightning from a cloud: The magnetic field would be too small, too small by orders of magnitude. Uhlenbeck asked Ehrenfest not to submit the paper, but it was too late. The paper was published, and to this day is considered an important, groundbreaking work. But the fact that the electron cannot spin fast enough is true, and we still live with this little embarrassment. When called to court, we swear that spin is a quantum mechanical phenomenon, and cannot be explained classically. (By the way, I retrieved some of this information from a transcription of a thoroughly delightful lecture given by Goudsmit to the Dutch Physical Society in April 1971.[3])

Spin is measured in units of Planck's constant divided by $2\pi$, or $\hbar$. For example, if you put an electron in a magnetic field, it can have only two orientations. Suppose the magnetic field is in the $z$ direction. Although spin is a vector, meaning it can point in any direction in three dimensions, we usually only care about the component that is in the direction of the field: the $z$ components. Calling this $S_z$, Goudsmit and Uhlenbeck deduced that the electron could have only two values for $S_z$, $+\hbar/2$ or $-\hbar/2$. We usually take the $\hbar$ for granted and do not even say it. For example, try this experiment.

"What is the spin of an electron?" you ask.

"One half," says any physicist, all the while knowing everything I just explained.

Try another experiment.

"What is the spin of a proton?" you ask.

"One half," says any physicist.

Do not give up; keep at it.

"What is the spin of a quark?" you ask.

"One half," says any physicist.

Do not forget about photons, and those Z-bosons that I mentioned.

"What is the spin of a photon?" you ask.

"One," says any physicist,

"What is the spin of a Z-boson?"

"One."

Okay, you see my point: There are two kinds of particles, spin 1/2 and spin 1. Actually, there can be multiples of these, so in addition to spin-1 particles we may have spin-2 particles, and so on, but there are only two genders: spin 1/2 and spin 1 (and the multiples). This is so important that we have names for these: Fermions are the half-integer spin, and bosons are the integer-spin particles.

If you think men and women are different, it's nothing compared to fermions and bosons. The biggest difference is that bosons like each other and fermions do not. What I mean is that you can put as many bosons as you like in the same quantum-mechanical state, but you can only put one fermion in each state. If this were not true, your chair would not hold you up, and you would sink to the center of the Earth along with everything else.

I do not want to belabor this, but I must belabor this. In building up atoms, each time you add an electron—a fermion— it cannot go into the same state as the previous electron, and therefore electrons build higher and higher energy states. If you squeeze two atoms together, the neighboring electrons cannot seep into filled states, which keeps them apart, which also keeps you and I above ground (so far). (This also explains what keeps white dwarfs from collapsing, and we call it *electron degeneracy*. It also explains what prevents neutron stars from collapsing—*neutron degeneracy*.)

To me, nothing is stranger in all of physics. Why are there two kinds of particles? And why are they so different? This bothered other people as well, and some of them opined, *hmm, maybe they are not really so different.*

Let's pretend you are a physicist. You have a great idea, and you work on it for years, and derive your commutation relations, like the ones we discussed earlier. They are wrong, however, because they have an added extra term proportional to (N-26), where N is the number of dimensions of space-time, which we know is four (three space, plus time).

**? Question:**

*What do you do?*

☐ 1. Throw up your hands and change occupations.

☐ 2. Admit your theory is wrong, and try again.

☐ 3. Assume your theory is correct, but that we really live in 26 dimensions, which zeros out the scofflaw term and makes your commutation relations correct.

This reminds me of Dirac, a little. Although his equations predicted a lot of things we knew were correct, he had trouble with the negative energy states. Instead of giving up he invented a new way to look at particles, and his ideas were subsequently used to view the positron. String theorists were not deterred easily, and they simply assumed we lived in 26 dimensions, and nature somehow hides 23 spatial dimensions from our view. Is this hubris or grand intuition? Some of both, I suspect.

To see how nature can deal such a sleight of hand, I think back to my days when I lived in Poughkeepsie, New York. I had a third-floor climb-up, with a wobbly deck big enough for one chair and one physicist. I could look down at a hose, winding across a small patch of weeds with a few blades of grass in between the dandelions. If you picture a bug crawling on the hose, you will give him two dimensions: north-south or east-west. But if you stroll out into the bed of poison ivy and peer more closely, you will see the bug merrily go round and round, up and down, happily circling his third dimension with insouciant abandon. In this case the third dimension is the hose, but, from the wobbly heights above, you cannot see it.

So living in 26 dimensions is not so bad if can figure a way to have 23 of them close up on themselves, on a scale so small that we have not been able to observe it. Unfortunately that was not the only problem with string theory; it was also a theory of bosons. If you have been reading this book you know right off the bat that this theory cannot describe nature. Fermions are to nature as light is to stars, and any theory without them is wrong. In physics when we come across this situation, but want to study it anyway, we sometimes call it a *toy model*, the idea being that playing with toys is a good thing.

In 1984, Ed Witton, John Schwarz, and Michael Green (some, but not all, of the main formulators of string theory), published their book on string theory. (It is green, by the way.) I delved into the tome one halcyon summer, merrily deriving equation after equation. And then came a brick wall. They said, "It may be shown that blah blah blah." I could *not* show that, and I went to the original literature and gave many of their papers a full exegesis, but I was stuck.

Physicists, by the way, never work on one project at a time, and I was funded by NASA to look into solar system tests of general relativity at the time. I also had a small Navy grant to look into magnetic field fluctuations and communications with submarines, when I was called over to the Office of Restricted Funds. To me, this is like being called to the Oracle at Delphi, and I could not comprehend their orphic speech. I made a mistake in the accounting system, which is arcane and a dozen times more difficult to understand than string theory. But it turned out to be good news: I had more money than I thought, and had to spend it within a year. Spending money is easy, so I hired a postdoc from MIT to give us a course in string theory. I told him the main thing I wanted to understand was the "blah blah blah."

The "blah blah blah" is of critical importance; it is the equations that represent the three fields I mentioned. It was the three fields that were present in my theory, and, when Hoseung Li finally derived them, I could barely move. Some years previous to this I had realized that a problem with my theory was that I had to modify the source so that it was something more than simply a point. I succeeded, but had trouble understanding the physical significance of what I had done. After Li's derivation, I went back to my theory and modified it by assuming the source was a string, and everything worked out just right.

Times like these are heaven for a physicist. I believed I understood nature, at least on this point, better and deeper than anyone else on the planet. The feeling was so strong I was not in a rush to publish my latest results. But when I did it was disappointing: I received a few invitations to give talks but interest faded fast, especially in the United States.

Back to string theory. Who can take a model seriously if it does not allow for fermions? No one. What's worse, it predicts tachyons, particles that travel faster than the speed of light. So, the next task was to introduce fermions into the theory, and it was achieved by using an idea I mentioned earlier: supersymmetry. Supersymmetry is a symmetry between fermions and bosons. It contains operators that are heretical: They transform bosons to fermions, and vice versa. Nowadays when we talk about string theory, we are really referring to superstring theory.

It turns out that superstring theory, similar to its progenitor bosonic string theory, cannot exist in four dimensions, but we do find an improvement: It can live in 10 dimensions. If somebody can come up with a natural and convincing way for "spontaneous" compaction of the other six dimensions, we would probably be fighting another string theory revolution.

Here is a conversation between Professor String and Professor Point.

"Surely you are joking, Professor String; everything you predict is wrong."

"My dear Professor Point, do not be so narrow-minded."

"You predict that the photon has a superpartner, the photino, but we have never seen it."

"That does not mean it does not exist."

"What about the squark, selectron, and Higgsino, not to mention the neutralino? All these superpartners, they have never been observed."

"That does not mean they do not exist."

"But Professor String, your theory predicts nine spatial dimensions. I only see three; there is no evidence for six more."

"You must think outside the point. There may be effects of these dimensions of which we are not aware."

"Come on, you cannot even predict a simple cross-section."

"Perhaps, but string theory contains the Standard Model, and the Standard Model predicts all kinds of things we have measured."

You heard the old saw: If you put enough monkeys behind enough word processors, you will get this book. If you put enough physicists in the same room, you will get this dialogue. Professor Point mentioned some superpartners, and an especially intriguing one is the neutralino, which, like all superpartners, is fun to say. Let me first remind you of the three generations (families) of quarks. There is the electron, the heavy electron called the *muon*, and the most massive of all: the tau particle. These fermions are identical except for mass, and the tau likes to decay to lower mass particles, and so does the muon, which is why they are so scarce. The electron is stuck, it is the lightest of this class, and with nothing lighter to decay into, is doomed to a life without end (unless a positron or something equally life-threatening comes along). This is why the number of electrons in the smallest bacteria is bigger than the number of people on Earth.

The neutralino is similar to the electron: the lightest in its class. It is like a combo meal, made from different superpartners, and is an example of cold dark matter. Neutralinos are expected to weigh in at 10 to 1,000 times the mass of a proton. With nothing to decay into, the universe may be filled with neutralinos, but where oh where can they be? Everywhere!

In Chapter 2, I forewarned of another kind of substance that might account for dark matter, and this is it. Neutralinos may have been created in the first febrile moments of existence, when the raging universe had more energy than it knew what to do with. As the universe cooled, these relic particles adopted a cloistered existence, living in the cold halos of galaxies.

A prediction such as this makes the theory exciting, but another issue casts a shadow over the theory: There are *five* string theories. One theory contains open and closed (loops) strings, others only closed strings, and, in addition, the symmetry groups differ. Physics is an experimental science, and, without hard predictions that can be measured, we are unable to know which theory, if any, is correct.

This forces me into a brief detour that takes us back to general relativity. Buoyed by the success of going from three to four dimensions, Polish mathematician Theodor Kaluza mused, "Maybe we should consider five-dimensional space-time."

Kaluza found he could unify gravitation and electromagnetism into a single theory by assuming we live in a five-dimensional world (four-plus-one, as we say, to emphasize that one dimension is time). Oscar Klein made the idea plausible by assuming the fifth dimension was closed up on itself, and small.

I came across this theory when I was a graduate student, and toiled through my second summer re-deriving everything, and trying out a few ideas of my own, but my advisor told me that this, like the Model-T, is a thing of the past. But, with the requirement of 10 dimensions in string theory, Kaluza's idea flared back to life like a nova.

My point is that when a new idea comes along, push it until it breaks. Going from a point to string generated intense interest, so why not go from a string to a sheet, or from a sheet to blob? We call a sheet a *membrane*, or, more specifically, a 2-brane, whereas the string is a 1-brane and the point is a 0-brane. In general, the idea is to think about p-branes, where p is any integer less than 10 (in 10 dimensions). And no, I did not make up this notation.

Perhaps it looks more complicated than ever, with strings, branes, and five different string theories swimming through murky seas, but the water seemed to clear when dualities were discovered. A duality is a transformation from one string theory to another, transforming, for example, from a case in which one theory is describing high-momentum strings to a situation in which it has high tension (tension arises from stretching the string, just like an elastic band). These ideas led to the second string theory revolution, or M-theory, developed by Ed Witten in 1995. What does the M stand for? People have attributed everything from Mother to Mud, depending on if you are pro or con. I should add that there are not a lot of the former advocates at this point.

Are string theory and M-theory just a lot of interesting math with little or no connection to our world? I do not know, but it reminds me of the early 1920s. Many physicists thought of general relativity as complex math, something much too knotty to describe our world. But as the theory passed test after test it was finally graduated with honors, and joined in the working ranks of our tools of understanding.

String theory has a long way to go.

*– Chapter Nine –*

# Origin of the Universe

And in the beginning there was nothing, not even darkness. And one aleatory moment later there was a great, numinous ball of energy, and as the great ball grew, it cooled, forming matter like morning's dew. And a billion years later galaxies were strewn across the new blackness like fireflies at night. And their radiance spread across the vastness, seeding new galaxies and stars like children. And soon we arrived, and we understand all this. And this is our universe.

The second-biggest mistake you can make about space is to imagine an infinite, black, empty space, when, all of a sudden, there is an explosion—the big bang (the first mistake is not to imagine at all). The reason this is wrong is that before the big bang there *was* no space, and before the big bang there *was* no time. Before the big bang there was *nothing*. The big bang created space, and the big bang created time. And the big bang created everything that was in it.

Not only is this event impossible to imagine, but it is also impossible to describe. We do not have the physics to correctly explain the first instants of our universe—but we can try, hopefully doing better in the future. This is yet another reason why I call this book *The Unknown Universe*.

If, at some time, there was nothing—no space, no time, no mass—then what caused the change? There was nothing to make the change. Plus, the first sentence of this paragraph makes no sense: Because time did not exist before the big bang, I should not write "at some time." What came before the big bang? The word *before* means "occur earlier in time." Therefore, because time did not exist before the big bang, there *was* no "before the big bang." All the mass (we think of mass as a form of energy, or vice versa) of all the galaxies was created in the instant of big bang, but where did it come from? One of our most cherished principles is conservation of energy. After the big bang we have an enormous amount of energy that did not exist until the big bang occurred. Was energy conserved?

It is not time to throw our hands up in despair. We'll take the big bang as a given and start to apply our physics as best we can after this momentous occasion. Let us start in the 19th century, when people were increasingly interested in using science to explain the world in which we live.

Every once in while someone pops a great question, the kind that makes you smile and say, "I never thought about that." One of these is the old standby, "Why is the sky blue?"

The first time you hear it, it is great. Let me give you the brief answer and ask you another question: Why do you even see the sky in the first place—blue, pink, or green?

Most of the light from the sun passes through the atmosphere, but some of it is scattered. When we say *scattered*, we are thinking of an entire quantum-mechanical process that boils down to this: Some atoms and molecules in the atmosphere absorb light from the sun, and then re-emit light. The reemission goes in all directions. This is why we see anything at all; otherwise, the sky would be as black as night. But why is it blue? Because the shorter wavelength (blue) is "scattered" more than red. Therefore, when you look up and see the scattered light, it is blue because most of the red part of the spectrum passed right through with little scattering, while the blue was courteous enough interact with the atoms aloft. (By the way, this also explains why sunsets are red. When you look toward the sun, most of the blue has been scattered away, leaving only red.) So, the sky is blue because sunsets are red, and sunsets are red because the sky is blue.

Now let us float back to the 19th century, when people knew about Newton's law of gravitation, and also knew there were stars as far as the eye could see. Then why, they wondered, didn't everything collapse to the center? The only answer was that it was infinite and uniform, so that, no matter where in the universe you stand, there is an equal amount of matter to your right as to your left. Two equal and opposite forces canceling, giving us the static universe. This was the prevailing view, an infinite static universe, as old as the hills.

This was the view when Beethoven completed what he considered his greatest achievement: the *Missa Solemnis*. It was 1823, and German astronomer Heinrich Olbers was already suspicious of this prevailing view of the unknown universe.

"Why is the night sky black?" asks Heinrich.

"You are ignoring moonlight?" asks Ludwig.

"Yes, I am asking about the sky. Why is it black?"

"Because the sun has set," replies Ludwig, who immediately goes back to composing.

To see why Beethoven should have stuck to music, here is Olbers's paradox. I remind you it is based on the notion that the universe is infinite and uniform. Olbers envisioned sectioning the universe in annuli (an annulus is a ring-shaped shell, such as we find in onions) of ever-increasing radius. Although the stars in the outer shells are more faint (because they are more distant), there are more stars in them, because those shells are thicker. In fact, Olbers adjusted the increasing radius of the annuli so that they each contained the same amount of light. When you add up all these shells, you get infinity, so that the night sky should be very bright indeed.

A few things got me interested in physics when I was still too young to spell it, and Olbers's paradox was one of them.

"Well, maybe the universe is not infinite?" tries Ludwig.

"Then it would collapse to the center," says Heinrich.

"I should imagine, therefore," continues the great composer, "that these outer shells of yours are not really so dense as you imagine, and in fact they become less dense, like the pianissimo notes of a flute."

"Then again, my dear Ludwig, it would collapse to the center."

Of course I have no idea if these two men ever met, but this dialogue, minus the bad moonlight joke and the dig against the flute, could have been with anyone.

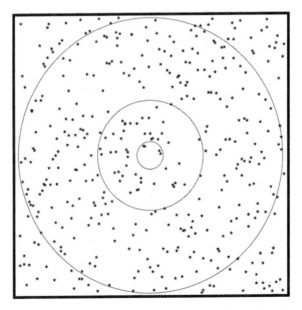

*Figure 9.1. A random, but, on average, uniform distribution of stars.*

No one could shed light on Olbers's paradox for the rest of the century. It went unsolved until the discoveries of Slipher and Hubble I described in Chapter 1, and the subsequent expanding universe we now accept.

Another thing about physics that has always interested me is how easy it becomes on Monday morning. Rephrasing Olbers, we can say, if the night sky is black, then the universe is not infinite, static, and uniform. Bingo! Why didn't anybody get this sooner? If a miracle occurs, and a copy of this book survives into future centuries, and someone reads it, surely he or she will smile and say, "Those poor ignorant people, how little they knew." But I digress.

We can never imagine how hard it was for people to adopt the view of an expanding universe. Olbers's little problem might have been solved by this enlightened view, and Einstein's confrontation with the failure of his (original) equations could have been overcome by it, but no one had the imagination or foresight to make a convincing argument that this was so. Hubble's hard data was another thing, and the window opened. I wonder how many other problems such as this are staring us down, as we refuse to open our eyes.

In your mind, reverse the direction of time and think about the universe. Everything is now contracting, and long after civilizations unwind we see great clusters of galaxies fall together. Eventually we envision everything crashing together, creating unimaginably high temperatures at which atoms cannot exist, and soon nuclei are burst asunder and crushed like grapes in a winery. And as we collapse even further, and all of the mass and energy of the universe is inside the size of an apple, not even neutrons and protons exist, and there is a superdense collection of only a handful of kinds of particles, including quarks, electrons, and exchange particles. And then there is nothing, nothing at all.

Because your mind is so agile, I will ask you stop just before everything reaches a point, and once again reverse the direction of time, so that we have an expanding universe. The shortest time is often taken to be the Planck time: $T_p = 10^{-43}$ seconds. We believe that at times shorter than the Planck time, and distances shorter than the Planck length, it is too risky to make predictions about anything, because space-time itself is as ragged as an old fisherman's net. So let us start there: The universe is $T_p$ seconds old and expanding fast.[1]

If supersymmetry is correct, all of the forces are of the same strength at this early time. This should sound strange to you: How were they equal then if they are so disparate now? We think of the strength of interactions in terms of coupling constants, but do not be fooled by the term *constant*. At high energies, the strength of the interaction changes. We can interpret this by saying that the coupling constants change. The early universe represents the most energetic time of all, which explains why the world is egalitarian. As it expands, however, the heady times of true democracy cool down—the energy is lowered—and what we see today is the low-energy limit, in which all the symmetries are broken and the forces are as different as Jupiter and Pluto.

The next stage requires a slight detour into the realm of phase changes. Many we see here on Earth, such as the lake surface succumbing to January's cold breath, or feeling the cooling powers of sweat as it evaporates from our bodies as we struggle on the tennis court. Phase changes also involve a lot of energy: The ice will take all of March's warmth to melt.

Phase changes can also be violent, to which I can personally attest due to my tea-drinking phase. I put a clean cup of water in the microwave and made it good and hot. I dropped in loose tea leaves, and, like a trick of Merlin, the water exploded out of the cup, sending leaves across the room like the eruption of Mt. St. Helens. (I learned my lesson: Stick to coffee.) The water became superheated, slightly above the famous 212 degrees, but did not boil because there were no centers to initiate the phase change (liquid to gas). I changed all that when I threw in the leaves.

In the early universe we suspect there was a phase change. No one was making tea, but a field—sometimes called the *inflaton*—was changing phase, and, like my water, giving off energy. A lot of energy. It caused the space to explode, expanding exponentially, and in a flash the universe grew by an unimaginable amount. This brief period ended when the universe was $T_p$ seconds old and, as coined in the 1970s, is called the *inflationary period*.

In these first instants, we say that the universe is radiation-dominated, as opposed to our current matter-dominated era. In the early stages particles are allowed to form, but before they know what hits them, the extreme density and high temperature will cause them to be annihilated, meaning converted into photons. The universe is essentially a ball of light, brighter than anything we can imagine.

The universe is expanding at practically the speed of light, which means it is cooling quickly. For a radiation-dominated universe, *cooling* means that the photons are becoming lower-energy photons. From a classical point of view, the expanding space causes the wavelength of the light to increase, which is equivalent to causing the frequency to decrease. This implies lower-energy photons.

As the universe continues to cool, the symmetry of the forces begins to break, and gravity emerges as the force we now know it, and the Standard Model of physics begins to kick in. If supersymmetry is correct, many exotic particles of energy 1 TeV or so could have been created. The stable particles may hang around to this day, and account for such things as dark matter, but this is speculation fueled by the dark matter issue I described in Chapter 2.

By the time the universe is a microsecond old, it consists of quarks, gluons, and photons. Other objects may be created, but their lifetime is short. After we reach a few seconds of age, protons and neutrons form, and after a minute or two the first nuclei form. After another minute or two, the universe is too cold to produce fusion, and calculations show we have a universe filled with 75 percent hydrogen and about 25 percent helium. Like pepper in a stew, there are a few other nuclei around, but they do not add much nutrition. The hydrogen and helium interact viciously with the photons during this time.

Matter domination occurs at 70,000 years. After this, there is more matter than radiation, but at about 400,000 years a major change occurs: combination. The atoms become stable, independent of the photons, which until then were continually breaking apart the atoms. This is when the universe becomes "clear," more similar to what we see now, except there is no structure. (All of the literature calls this phase *recombination*, but because stable atoms form for the first time, *combination* seems better.)

Structure probably begins just after the matter domination, when random fluctuations in the density begin to snowball, forming the seeds of the earliest galaxies (it is believed that dark matter has a hand in galaxy formation). After this, gravity dominates the universe, and, as the years and centuries go by, the universe grows as the inchoate density fluctuations collapse into galaxies and stars. By the time the universe is one billion years old, it is strewn with galaxies and stars, and by the time the universe is 14 billion years old, its inhabitants have acquired the intelligence to see this marvel.

After a year of preparation, astronomers turned the Hubble Space Telescope toward a dark part of the sky, away from the Milky Way, toward the handle of the Big Dipper. For 10 days in December of 1995, the great telescope took pictures continuously, with exposure time between 15 and 40 minutes. Combining 342 exposures in the infrared, red, blue, and ultraviolet, we were able to see a typical piece of the universe as it was 13 or so billion years ago, when it was one billion years old. This is called the Hubble Deep Space shot, and represents one of the earliest views of the universe we have (and is my personal favorite of all cosmic pictures).[2]

About four billion years after this, our solar system was formed, using bits and pieces of the remnants of an untold number of supernova that gave us the iron and other heavy elements we find on Earth. In another billion or so years life began on this planet, and nearly four billion years after that I was a graduate student, arguing about the future of our universe. I cannot convey how amazed I am that the universe evolved to what we see today.

Up until a few years ago, we could not decide on the ultimate fate of our universe. According to Einstein's equations, two futures were possible, similar to the two choices of what could happen when you threw that rock up from the moon's surface: It can continue to expand forever, or it can reach a maximum size and collapse. Which of the two scenarios plays out depends on the density of matter, which is tantalizingly close to the critical value lying on the fence separating the two possibilities. Every new cosmological observation was greeted with great anticipation, and we eagerly looked for more evidence. It seems the ultimate fate of our universe became a question that burned nearly as hot as the one concerning its origin.

It was answered recently, as explained in Chapter 2. We now have reason to believe that the universe is not slowing down, but, like the moon rock that turned into a rocket and started speeding up, the expansion of the universe is accelerating.

The idea behind all of our theory and observations, calculations and experiments, is science, and that we can do better than the ancients who created gods for unknowns. Nevertheless there are those who eschew science for faith, and the abiding debate concerning evolution versus creationism rages like a wildfire, sometimes smoldering, sometimes whipped to an inferno, fueled by political winds at gale force. The dispute knifes through coffee shops and schools, from workplaces to Washington, cutting a divide between political parties and families alike.

The two views certainly seem to be diametrically opposed to each other, and although some scholars are able to find reconciliation through a liberal interpretation of scriptures, I would like to point out that science cannot rule out creationism. It is rather an obvious point, but has some interesting ramifications: Science only shows us what is simpler.

For example, the world could have been created five minutes ago. In this case the Creator would have had to put a lot of thought and care into the job. The Creator would have created your clothes, your memory of buying those clothes, along with the myriad bones and fossils spread across the continents, with just the right ratio of carbon-12 to carbon-14 (carbon dating), and even the intriguing DNA sequences that indicate the migration of human life from Africa. Distant stars, as well as their radiance that fills the cosmos, were also created five minutes ago, and all those other things you might imagine.

On the other hand, if the universe were created 14 billion years ago, all the Creator had to do was make a bunch of hydrogen—the simplest atom—and a little helium, and sit back and wait, letting all those fossils and distant stars worm their way into the present day naturally. And of course, a job as easy as this needs no Creator.

There is no experiment, no observation, no evidence that can disprove that the world was created five minutes ago. To science, it just seems unnatural, or too difficult. (If the Creator is omnipotent, then who are we to judge what is difficult and what is easy? But that is not science; that is belief.)

Nevertheless, it is interesting to pursue the idea that the world was created five minutes ago. It points to a more humane and talented Creator. All the recorded suffering, from Stalin's despicable mass-murder campaign to the numerous religious wars, never happened. The world was created with this history, perhaps in an attempt to steer us away from future horrors while endowing us with free will.

The wonderful music of Bach, Mozart, and Beethoven, and so many other composers, sings louder now than ever before. *La Bohemme* has never been performed, but what a treat we are in for. The Creator was musical, artistic, and a scientific genius. The statue of David, Beethoven's *Ninth Symphony*, and the theory of relativity were made by the artist-composer-physicist-writer, the Creator. The Creator also gave us a multitude of religions and devils, heroes and heels, rich and poor, and spread evidence of a few giants across the globe like nuggets of gold.

We look back on the ancients with quaint tolerance, intrigued but not converted by their beliefs that gods bred with humans, and gold-filled coffers helped the dead galumph through the afterlife. Certainly we will be viewed under a similar light, someday, when this pesky question haunting us will only echo from the distant cliff of history. But do not take this message too lightly; it may have been written by the Creator. That is not science, that is belief. From now on, I will stick to science, so let's get back to business.

I would like to emphasize that our universe is an example of curved space-time, which brings up the parable of Bill and Hillary. Two explorers are stranded on a world that is a perfect sphere, covered with sand. They are standing on the equator and get into an argument about the best course of action. Unable to agree, Hillary goes west by north, exactly 10 degrees from the equator. Bill goes west by south, also exactly 10 degrees from the equator. They agree to go in absolutely straight lines, which they measure by making straight lines (180 degrees) in the sand. They also agree to travel at precisely the same speed.

"Are you sure about the straight lines, Hillary?"

"They are geodesics, Bill—the shortest distance between two points."

As they depart, they watch each other fade into the distance, but keep to their agreement. They trudge along day after day, week after week, until they lose all track of time, until Bill sees a mirage to his north, another lone traveler. His mind plays tricks on him, and he begins to think it is Hillary. When they meet they embrace, apologizing for their alleged misdeeds.

"Hillary, I am so glad you changed your mind, and changed your course."

"I did no such thing, Bill. I remained on a course straight and true."

"Then how did we meet?"

"Curved space, Bill, curved space."

The two-dimensional surface of a sphere is a two-dimensional curved space. Because we cannot visualize curved three-dimensional space, we must either resort to mathematics or use a lower-dimensional space that we can envision.

When I taught astronomy another favorite demo was the pennies on the balloon. I would get a balloon that could be blown up to about 5 feet in diameter. I would then glue on some pennies, and, using a marker, draw some circles the size of the pennies. I would deflate the balloon and take it (along with an electric pump) to class, announcing that today I will demonstrate the big bang.

It is always dangerous when a theorist begins to tinker with real things. It was hard to find the right kind of glue, and on one occasion, after the glue ate through the rubber of the balloon, a penny shot out like a bullet, making a home in the hair of one of my students. Another time I used the wrong marker, and when the universe reached a measly three feet, it burst asunder, scaring the heck out me but bringing great delight to the class.

After a few years I had the right stuff, and blew the balloon up, reminding the students to watch what happens to the pennies. Although they are at rest with respect to the rubber, the distance between them grows. This happens not because they are moving but because space is expanding. Hubble's law falls out of the solutions to Einstein's equations as a result of this interpretation.

**Question:**

*Should the galaxies be represented by the pennies, or the circles I made with the marker?*

☐ 1. The marker circles. After all, if space is expanding, it is expanding everywhere.

☐ 2. The pennies. The galaxies are gravitationally bound; this force holds them together.

The answer is 2, although it may not seem obvious. To obtain the expanding-universe solution, we assume that matter density is a fluid, like water, or a gas, uniform and evenly distributed. All of space would expand if this were the true distribution of matter, but it is not.

Earlier I told you about Einstein's "biggest blunder" (his words, not mine): the cosmological constant. As I explained, he invented it in order to find cosmological solutions to his equation that yield a static universe, similar to the one that rattled Olber so much. But while Einstein was changing his equations to fit his view of the universe, others were following his equations to see the universe in its true light.

By 1924, Russian mathematician Aleksander Friedman discovered solutions to Einstein's equations that predicted expansion. This occurred a few years before Hubble's famous revelations. Not only did he predict expansion, but his solutions also showed there must have been a big bang.

During this time Belgian Georges Abbe Lamaitre was ordained a Roman Catholic Priest, but he was not content with what he read in the scriptures, unless you add Einstein's work to the canon. He also discovered the solutions of Aleksander Friedman, and in so doing—with the cold logic of a mathematician—was able to see creation from a different perch than most of his brethren. Another two seers to see the exploding universe were Robertson and Walker, and the standard cosmological model we now have is usually referred to as the Friedman Robertson Walker cosmology. In the trenches, we usually call the geometry the Robertson Walker space-time, and, when we assume that matter is distributed uniformly through the space, we call this the Friedman universe.

By the way, this, to me, is the greatest honor of all—having the universe named after you. Some people get a law, like Torricelli's law that describes the water coming out of a bucket with a hole in it, or a mathematical law like Gauss's law, or even a set of equations, like Maxwell's equations that are the laws of electromagnetism. But the whole shimola? Nothing is bigger than that.

There have been other attempts and interpretations of our universe. An intriguing solution was found in 1917 by Dutchman William de Sitter. It presents an empty static universe, but is nevertheless curved space-time. Even more interesting, the farther away an object is, the more it is redshifted. Einstein did not like this at all, because he believed it is the distribution of matter that determines the structure of space.

A very modern view of our universe interprets the big bang differently. The universe might be infinitely old, and long ago was collapsing. When it became very small and extremely hot, some other process kicked in, such as inflation, or quantum repulsion, that caused the universe to "bounce," and it has been expanding ever since.

A helpful way to understand cosmology is to watch someone ride a bicycle at night. If he or she has a reflector mounted on the tire it sketches out an interesting pattern, shown in Figure 9.2, which is known as a cycloid.

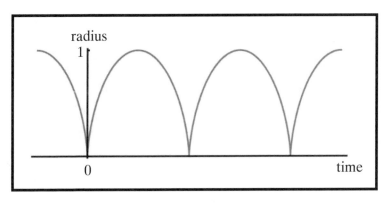

*Figure 9.2. The mathematic curve called a cycloid describes the radius of the universe as a function of time, but the zero-radius points are a problem.*

This graph also represents the radius of the universe as a function of time, which is true if we resort to using the Friedman Roberston Walker model. On the vertical axis is the radius divided by the maximum radius (so that the maximum value is 1), and on the horizontal axis I plot time. The actual numbers depend on the density of the universe, and using the observed "best estimate" for the density this can be a reasonable approximation of our universe, which is represented by only a single cycle of this graph.

The most striking features are the points along the time axis, when the radius is zero. Some old-timers interpreted this as a continually bouncing universe, but nowadays we reject this notion, noting that, as the radius approaches 0, the model fails. It fails partly due to the fact that we must account for radiation pressure, not to mention inflation, but mostly because we know, for small lengths, we need quantum gravity. Nevertheless, the idea, as do all intriguing ideas, stubbornly lives on. Richard Tolman, who wrote a beautiful text on general relativity in the 1930s, concluded, on thermodynamic grounds, that the cycles must become shorter as you go back in time, so we still end up with an initial singularity. The ultimate kibosh struck later, when singularity theorems showed that, once contraction starts, the final-crunch singularity is sure as taxes. If that were not enough, as I explained in Chapter 1, the expansion is accelerating, not slowing down to a contraction phase. But theorems are based on assumptions and observations, and sometimes observations can by as reliable as a bribed witness. The cyclic universe is back, at least in theory. If we live in a brave new world, in other words, in the fiefdom of strings, the acceleration we see may be only part of the picture, and other dimensions may be contracting, like a balloon being squeezed at one end while bulging out on the other side. Perhaps the squeezing/pulling structure is following some natural period of oscillation, like a great cosmic pendulum. A wonderful account of these ideas is given in *Endless Universe*.[3]

And this brings our unknown universe to the forefront. It could be one of human's greatest achievements, understanding, and maybe predicting, the beginning of everything, but we cannot. We cannot understand the early times of our universe and we cannot understand quantum gravity, two of the biggest conundrums we face.

Perhaps you will allow me to hop up on the soap box one more time. In the 1940s Vanevar Bush, dean of engineering at M.I.T., enunciated what became know as the Bush Doctrine. It states that funding science is good because economic prosperity follows. This has been quoted like scripture since the 1960s, when it seemed to be an accurate prophecy, and is used to this day to justify research. But I have always found this most depressing. I would argue that funding science is good because understanding the universe in which we live is good.

But that's me.

– *Chapter Ten* –
# Mysterinos

Just for the record, this is another word I made up: *mysterino*. It arises from the Fermi coinage of *neutrino* from *neutron* and blends into the proliferation of "ino" particles from supersymmetry, such as the neutralino or photino. It means "little mystery," or something we have not seen.

I will begin with antimatter, first discovered by Carl Anderson, as I described in Chapter 8. To understand the physics of elementary particles we turn to Terry Bradshaw, ex-quarterback of the Pittsburgh Steelers. When a football is thrown well, it spins along its axis. Suppose a right-handed QB tosses the ball downfield. He watches the ball go away from him, spinning clockwise. We (physicists) say that the ball has positive helicity. Most QBs were right-handed, and young upstart left-handed QBs were something to be wary of. After all, the receivers are used to catching positive helicity passes, they moaned (in different words). A left-hander will give the ball the wrong spin, and receivers will not be able to catch it. Bradshaw proved this statement to be as specious as it sounds, but when he threw a pass, the ball did spin the opposite way, and we say it has negative helicity.

Particles are nature's version of the poorly thrown pass. They spin off their axis, and, as they zoom from QB to receiver (from emitter to absorber), they are allowed to have either positive or negative helicity. Now consider the particle that Carl Anderson found in his cloud chamber: the positron (antielectron). If we slam two photons against each other, and if they have enough energy, they can create an electron-positron pair. The net charge is 0, because the positron has exactly the opposite charge of the electron, so charge is conserved. Angular momentum is also conserved because the positron will have spin opposite to the electron. And so there are antiprotons, antineutrons, antiquarks, and so on. When they meet, they annihilate each other to form photons. It is the most efficient source of energy in the universe. A big commotion occurred a few years ago when a scientist created anti-hydrogen, an atom consisting of an antiproton and a positron, but this is as far as we got. There are no antidogs or anticats, at least on Earth.

## Axion

Often in physics, a problem is really a solution in disguise. To explain, let us delve into the religion of particle physics, in which the holy trinity is charge conjugation (C), parity (P), and time reversal (T). C changes the sign of all charges, P is the mirror image, and T changes the sign of time, making time go backward. I still remember my graduate student days: I am taking the exam for a course in quantum field theory. There is only one exam, at the end, and it is oral, and I am scared. As I answer questions, however, I gain confidence rapidly, until I am asked to derive the CPT theorem.

I will not bother you with the gory details, but the CPT theorem states that if you perform all three operations (C, P, and T), any physical theory will not change. It is a cool theorem, and, if it is wrong, some of our most cherished notions must be thrown out the window. In the old days, before 1955, many physicists thought each was conserved separately (I am told). But the decay of a cobalt nucleus showed that P is not a true symmetry of nature, and later it was found that the weak force violates CP, the combined operation of C and P.

However, the strong nuclear force respects CP, and people, as with the famous pendulum, have swung the other way, and ask, *why*? The theory would be just as happy to violate CP as not, which reminds me of a fundamental rule. It was not until I discovered this that I was able to understand anything about particle physics. Basically it says that if anything can happen, it will happen. The converse is that if something does not happen, it is because Cerberus is standing guard, preventing it from going forward. For example, muons decay into lighter particles, and so do tau particles, but electrons do not because there is nothing lighter (and charged). In this case Cerberus is conservation of energy and charge, locking the electron into a long and happy life.

It is this spirit that brings the axion to life. If the strong interactions are allowed to violate CP, then they should. If CP is not violated, there must be reason (as modern thinking goes), and in this case Cerberus is the axion. The theory of the strong force can be tinkered with, and a term can be added that guards against CP violation. This term is what gives rise to the axion, and, as I mentioned earlier, we have been searching for this elusive particle for decades. If it exists, it may solve two problems: the lack of CP violation and the existence of dark matter.

## Tachyons

According to Einstein's special theory of relativity (1905), no particle can be accelerated to the speed light. At first glance this seems to limit us to a tiny speck of our universe, especially when you consider that it takes 100,000 years for light to eke its way across the Milky Way. Unlike many of my colleagues, I always looked at this as good news: It keeps the little green people far away, and we do not have to worry about Gort rampaging through our Capitol. But before I wander too far off course, let me get back to tachyons.

If you push on a ball (as you do when you throw it), you do work. Your energy is transformed into the kinetic energy of the ball. The faster the ball travels, the more kinetic energy it has. Before 1905 we had a formula for kinetic energy, but after 1905 we had the correct one. They both agree for low speeds, but, when you start knocking on light speed's door, they are like night and day. Here is a graph of the kinetic energy of an electron.

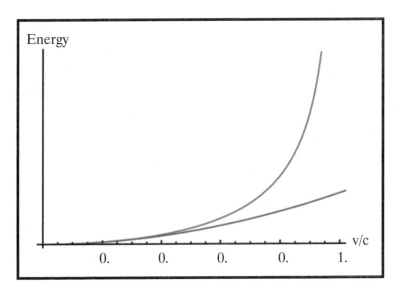

*Figure 10.1. The kinetic energy as a function of time. The lower line is before the theory of relativity, and the top line shows the correct function, indicating that no particle can be accelerated to the speed of light.*

The velocity is measured relative to the speed of light, $c$, and the energy is given in terms of the rest energy of the electron. People took the lower line for granted until 1905, Remember: The value of 1 on the horizontal axis is the speed of light, and people dreamed that, the more you push, the faster it goes, and could go well beyond the speed of light. But this not true: It is the result of the special theory of relativity. The more you push, the more energy you give the particle, but, as the graph indicates, the particle will never reach the speed of light, because that would take an infinite amount of energy.

You might infer that you can never make a particle that motors along faster than the speed of light, but this would be a hasty decision. Particles are created left and right, and there is nothing in this argument that prevents particles from being created with speeds greater than $c$. These hypothetical particles are tachyons, they can be anything from electrons to stringy objects, and they zip along at superluminal speeds.

A few other issues cause us to doubt their existence—the biggest, perhaps, being that they have never been observed. Also, we must tinker around with the formulas of special relativity, changing them from what Einstein derived, from equations that have been tested. Finally, there are issues with causality, but, discounting a few people I meet on the roads, no one has ever observed anything traveling faster than the speed of light, so tachyons are more of a mathematical curiosity than anything else. The worst part is, they demote you and me to the status of *tardyons*.

# Where Have All the Antiparticles Gone?

Two photons can mash together, annihilating each other, but what happens to conservation of energy? Nothing. A particle and its antiparticle can be formed, and, thanks to $E = mc^2$, energy is conserved. In fact, this happens all the time, but why particle-antiparticle pairs? Because nature conserves other things too, like charge and spin. Particle-antiparticle pairs conserve everything, so they are created, as I said, left and right. Now, back to the early universe, when it was so hot that photons ruled supreme. Particle-antiparticle pairs could be created, but they would also annihilate. As the universe continues to cool, from expansion, these pairs should begin to live longer, and by the time combination (recombination) rolls around, there should be as many particles as antiparticles. But this is not what we see; we see only particles, so where have all the antiparticles gone?

This is a real toe-tickler, and many physicists kick off their sandals and try to explain it. It is possible that there are vast regions of antiparticles, galaxies even, but it would seem that there would be matter-antimatter annihilation at the galactic edges, giving rise to telltale radiation—telltale radiation we have not observed.

We are in the courtyard of particles physicists, where symmetries rule like a judge. We might speculate, therefore, that there is a symmetry-breaking effect, some process that prefers matter over antimatter. For example, when we believed neutrinos were massless, we believed there was only one kind: left-handed neutrinos. We then built the Standard Model of elementary particles in such a way that only left-handed neutrinos are produced. If you accept the Standard Model as something basic, then we had an explanation of why there are no right-handed neutrinos. It is a circular argument, but it is something.

As far as the puzzle of missing antimatter, I am sorry to say I cannot give you the solution, and, although there a number of putative theories, this represents yet another piece of our wonderful but unknown universe.

## Negative Mass

Negative mass is nothing like antimatter, and we have never observed it. To describe it, by definition, we replace $m$ with $-m$, and from there we let the equations do the talking. One of the most important laws of all time is Newton's second, which states that the total force on a body is equal to its mass times its acceleration: $\mathbf{F} = m\mathbf{a}$. The boldface implies that those quantities, force and acceleration, are vectors, which means they have directions. The mass is not a vector (we call it a scalar), and Newton's second law, among other things, shows that the force and acceleration are in the same direction: If you push on a ball with a northward force, it will go north. When the ball hits the wall, the wall exerts a force opposing the motion (the "wall force" points south), so the ball accelerates in the opposite direction. But if the ball had negative mass, although the wall still exerts a southern force, the acceleration of the ball is opposite (that is what the negative sign does), so the ball accelerates north, through the wall. The harder the wall pushes, the greater the acceleration, so the ball must burst through the wall, speeding up as it does. I call this the armor-piercing effect: No matter how strong the armor is, the negative-mass projectile will break through it, gaining energy as it does.

Imagine two balls, one positive mass and one negative. A careful analysis using Newton's laws shows that the negative mass will chase the positive mass. Both energy and momentum are conserved, because one object has negative mass. This could explain why we do not see any, but now and then such a thing may zip past us.

## The Left-Handed Universe

Why is the speed of light $3 \times 10^{10}$ centimeters per second? Why is the electron mass $9.1 \times 10^{-28}$ grams? To me these are great mysteries, well above the level of mysterino status, but we have used these numbers so many times—in an untold number of calculations—we begin to take them for granted, and some of us lose our sophomoric sense of wonder. On the other hand, perhaps these numbers that haunt me are akin to the number that troubled Kepler, and, as with the number of planets, these numbers are just an accident of formation. The formation of what, I wonder...

As I explained earlier, electrons are like poorly thrown (wobbly) footballs, and they can be thrown from a right-hander (positive helicity) or a lefty (negative helicity). Same thing for neutrinos, if they have mass. Neutrino interactions are described by the Standard Model. However, only left-handed (negative helicity) neutrinos partake in the Standard Model. If right-handed neutrinos zip by, they would not feel the weak interaction and would appear completely immune to it attraction. For this reason, they are sometimes called *sterile neutrinos*.

In the early 1990s this gave me an idea. At the time, we believed that we were receiving only half the number of neutrinos the sun was emitting, and the solar neutrino problem raged on. I realized that torsion, part of the generalized theory of gravity, could interact with a massive neutrino, and I calculated the probability that it is flipped (sterilized, changed from left- to right-handed). I was very excited as I did this, merrily computing cross sections using the standard concepts of quantum mechanics, but, when I got the answer, my balloon popped. The probability of flipping turns out to be 1 in 10 million. The result was published anyway, and of course now we know the correct explanation (neutrino oscillations, which I described in Chapter 5).

This fact is not a prediction of the Standard Model. If it were, that would be a good thing. Instead, we write the interaction so that only left-handed neutrinos interact. Why is this? What is so special about left-handed neutrinos, and why are right-handed neutrinos bubonic, ostracized, with no form of interaction (except gravity)? I do not know, and this is very mysterious to me. I also think any theory worth its salt should explain such a thing.

But that's me.

## Gravitational Waves

These, above all other things we have never observed, are most believed in. Just the way Maxwell's equations predict the existence of electromagnetic waves, Einstein's equations predict the existence of gravitational waves. Because Einstein's theory has passed every classical test we have devised, this prediction cannot be wrong. In fact, they barely belong to this section of mysterinos, except that we have not yet detected them.

Similar to electromagnetic waves, they travel at the speed of light. They are created by an accelerating mass, such as a spinning neutron star, or two neutron stars colliding, or as stars collapse into a black holes. You can also create them here on Earth by waving your hand back and forth, but these would be much too weak to measure, which is why we do calculations describing the astronomical objects I just mentioned.

If a gravitational wave came directly at you, it would stretch and squeeze you sideways, and then elongate and shrink you along your length, but unless you are very close to a black hole you should not worry about these tiny effects. One class of gravitational wave–detectors consists of an aluminum alloy bar, a meter long by half a meter in diameter (or so). Along its length are piezoelectric crystals, which produce an electric current if they are squeezed. If a gravitational wave comes along, it will squeeze/elongate the bar, which in turn pinches the piezoelectrics together. The idea is to detect the current that the crystals produce. To be sure we do not accidentally measure a big old truck barreling down the road, two or more such bars are operated simultaneously, and we look for coincidence signals.

This experiment turns Michelson's oil drop experiment into a piece of cake. Calculations indicate that the strain (change in length divided by length), for the astronomical processes I mentioned is about $10^{-20}$. For a 1-meter length this implies that the change of length is $10^{-20}$ meters. Because an atom is not much bigger than $10^{-10}$ meters, this means we are looking for a change in length that is $10^{-10}$ (one 10-billionth) the size of an atom.

When I was a graduate student, excitement exploded through the physics community with the speed of light when it was announced that a gravitational wave had been measured. However, the announcement was premature, and subsequent analysis of the data showed the effect was an artifact, and not a real gravitational wave. Still, aluminum bars have been in continual operation ever since. All the additional information you might like to know can be found on the Website of the Department of Physics and Astronomy at Louisiana State University (*http://sam.phys.lsu.edu/*).

One day, when I was at the jet propulsion lab, I strolled across the campus and talked to Frank Estabrook. We discussed his calculations, and I asked him for a reprint. He opened a file cabinet and gingerly pulled out a thin, yellowed paper, only a few pages long.

"Be careful, Rich. This is my last copy."

Frank gave me his last reprint of a paper he published with Hugo Wahlquist that described how to measure gravitational waves using mirrors. The apparatus is an interferometer that capitalizes on the smallness of the wavelength of light. The larger the interferometer, the more sensitive it is, and in fact we hope that future large interferometers will provide yet another window to our universe by providing a new kind of telescope/observatory/antenna.

**Question:**

*To make an interferometer large enough to detect gravitational waves, and to provide a new "cosmic eyepiece," we plan to:*

☐　1.　Use a lab as big a gymnasium, making the interferometer more than 10 meters long.

❑  2.  Build underground tunnels, kilometers in length, to house the interferometer.

❑  3.  Launch a spacecraft that will deploy a bunch of satellites, each satellite acting as a mirror in the inteferometer.

The answer is all three! In fact, the first two are completed, and we are still planning the orbital interferometer LIGO, or Light Interferometric Gravitational wave Observatory.

# Fractional Dimensions

*This will really bake your noodle.*

—Oracle to Neo, *The Matrix*

Earlier I described the problem of renormalization, such that infinite quantities erupt like volcanoes, but their fire is tempered, and removed. There are different ways to accomplish this, and most techniques start by regularization. This means that the infinite part is separated from the "real" part, and then we apply a technique to kill off the infinity. One method is dimensional regularization. It turns out that we can relate the infinities to a function that becomes singular (infinite) at integer values (1, 2, 3, and so on). Suppose we call this function $\varepsilon$, which is very small, and pretend we live in $4 + \varepsilon$ dimensions. Then we can separate off the infinite part and then let $\varepsilon$ equal zero. (This is an old trick, showing how sneaky and clever physicists can be.)

I was a graduate student listening to a guest lecturer describe this in a seminar, which consisted of a dozen faculty and as many graduate students. One of the faculty, Professor Yergin, asked, "Why do you let $\varepsilon$ go to zero?" No one grasped the significance of his question right away, but, as he sat calmly waiting for an answer, I became very excited, and imagined some new possibilities I could barely comprehend. "Perhaps," Yergin continued, "we live in four-plus-epsilon dimensions, and this is what the mathematics is showing us. Perhaps you can determine epsilon."

Excited as I was, I was never able to get my arms around this concept, which sits out there, with so many other ideas that I shall revisit some day. Other people, however, have taken this more seriously, and derive geometrical relations in this orphic space where the number of dimensions has no respect for the integers, and can lie anywhere on the number line. And how would matter curve space-time in 3.141 dimensions? I do not know, but I just wanted to remind you that there are a lot more things in heaven and earth than Horatio's philosophy allows.

## *Pioneer 10*

In the summer of 1993 I found an atom-sized room in Pasadena for the summer. It was within walking distance of the Jet Propulsion Lab (if you have 45 minutes to spare), and the rent was enough to settle the national debt. From the JPL

I would take the commuter bus to the Cal Tech campus, and, as was my custom, hoof up seven flights of stairs to the seventh floor—the physics floor—of the library. This was back in the days when we held paper journals in our hands to read them. I was brushing up on modified theories of gravity that might explain the anomaly JPL was facing, but after several hours I pulled my nose out of the journals and took a peek outside. Everything had disappeared, and, as I stared into an impenetrable white haze, I thought of the early universe, before combination.

JPL's mission is unmanned flights, and I was there as a visiting scientist to use their data to help with solar system tests of general relativity. But, as I said, it was the summer of 1993, and the entire campus-like community was abuzz with talk of the *Mars Observer*. As I strolled across grass-lined walkways, amid a string quartet my friend Mashoon played in, and other people practicing to be statues, there were charts and billboards festooned throughout the campus, telling of the *Mars Observer*'s whereabouts, its capture orbit, and a lot of technical jargon that gets me and other physicists excited.

The ebullient atmosphere is redolent with anticipation, and, sitting next to a great fountain, I scribbled away, trying to solve the mystery that is *Pioneer 10*. Launched in 1972, it is now past Pluto, still emitting radio waves that are captured by the Deep Space Network (DSN), three large antennas scattered across the globe in the Mojave Desert,

Madrid, and Canberra. Expected to reach Aldebaran in some two million years, *Pioneer*'s feeble emission of not much more than a 100-watt bulb was received by the DSN back in 1993, and they also measured the change in wavelength. Using the Doppler effect (described in Chapter 1), the DSN was able to calculate its speed and acceleration, and this is where the head-scratching starts. I was working with John Anderson and Eunice Lau, who took the raw data and subtracted away the acceleration of every known force, from the solar wind to the effects of other planets. After this, the acceleration should be zero—but it is not. This is the problem: What mysterious force is pulling on *Pioneer 10*?

This tiny acceleration—$10^{-7}$ cm/s$^2$, or about one 10-billionth of the acceleration of an apple as it falls from the tree—has been the subject of many publications throughout the years, and it was great fun to be there, working on the problem as it unraveled. (For an introduction, check out *http://en.wikipedia.org/wiki/Pioneer_anomaly*.) John asked me to consider whether dust could account for the deceleration. This region of the solar system, beyond Pluto and between 40 and 60 AU, is called the Kuiper belt. It is loaded with primeval comets and debris, and could well be littered with a ring of dust.

A layer of dust would slow down *Pioneer* by friction, just as the atmosphere creates drag on projectiles. In fact, it is simple to derive the density of the putative dust to account for such a deceleration. However, that would also slow down the rate of rotation, and so I was able to rule out dust. I began looking at other theories of gravity, and the summer went by so quickly my wife had to call me and tell me to come home. But then the bomb was dropped. People kept a stiff upper lip for a day or two, but the JPL became more somber than a funeral home. The *Mars Observer* was lost and never found.

The origin of the anomalous acceleration of *Pioneer* remains as mysterious as the fate of the *Mars Observer*.

## Anthropic Principle

When I was a visiting assistant professor at Clemson University I went to the library every day. Although I would glance at a journal now and then, my driving passion stood behind the book checkout. With long, golden-blond hair and sparkling blue eyes, Nancy was the most beautiful woman I had ever seen. When I finally got the courage to speak I just blathered like Ralph Cramden. Somewhere in the mumbo jumbo that followed was a marriage proposal, but also a yes, and now, with offspring growing faster than kudzu, I look back in awe at the luck I had in Clemson.

I also stand, bowed and panting, on the tennis court as I watch my son rip a crosscourt winner, and ask, *what are the chances of that?* What if I did not accept that one-year visiting position at Clemson, and what if I broke left on the court just now instead of guessing right? I might have returned his shot.

And there are nearly an infinite number of other "what ifs," any one of which could have changed things enough to prevent that crosscourt winner, making it a miraculous event.

In fact, billions of billions of billions of things must have conspired just right, from the tension in his racket to the incomprehensible nuances of romance, for this event to occur. Confounded by these odds, instead of convincing myself that such a shot, by the odds, is impossible, I might counter with this argument: In reality, there are an untold number of universes—so many, in fact, that there is one for each of the various kinds of "what ifs" I just mentioned, and in most of these the crosscourt winner never occurred. I just happen to be in this one, where I am ambling over to the bench for water and oxygen. The idea of many worlds is not new, it goes back to our unfortunate friend Giordano Bruno, and to Epicurus, and once again has risen to popular speculation. I look at it as a physicist's version of Manifest Destiny.

What I am alluding to here is the anthropic principle. I personalized it a bit, but you can apply to the whole universe. If the mass of the helium nucleus were just a tad higher, then nuclear fusion (hydrogen to helium) would not occur. In this case stars, for a fleeting moment, would twinkle like fireflies (from the heat during collapse), but night would come quickly, a cold blackness scattered with unseen orbs of hydrogen.

In fact, if you begin to tweak almost any of the constants of nature, you would create a universe so much different from this one there is virtually no chance we would be here to ponder the issue. The fact that we are here, even though the odds are against it, implies there are many, many universes.

**Question:**
*What do you think?*

— ✳✳✳ —

# Epilogue

**July 11, 2150**

"My dear Professor Philomath, what in the world is in your hands?"

"It is an incunabulum, Professor Philophysics," he says, gently blowing some dust from the cover of the aged book. The ambient detectors signal excess aerosols and create a local low pressure nearby, quickly removing the ominous cloud.

"Wherever did you get such a thing?"

"Library archives. This is the last extant copy."

"*The Unknown Universe*? When was it published?"

"Long before we were born, Philomath. Look at the table of contents." They peer down, like Darwin bending over a new species. "Look at what they thought about cosmic rays!"

Finding something else risible they both chuckle as Philophysics says, "Look at this: dark matter."

"Oh my, they believed in all kinds of things!"

This specious statement brings on the cachinnations, and Philomath has to wipe a tear from his eye, but Philophysics gets serious and says,

"Philomath, look," his arcuate finger is skimming across the yellowed pages, tapping here and there. "These issues are still with us. Surely you are not a man who thinks all problems will be solved?"

"No, my dear Philophysics, but I am not a man; I am inevitability."

—— ✳✳✳ ——

# Notes

## Chapter 1
1. Physicists use what we call a *diffraction grating*, but it does essentially the same thing as a prism.
2. Max Planck said, "A new scientific theory does not triumph by convincing its opponents and making them see the light, but rather because its opponents die...." See Eisenstaedt, *Curious History*, 3.
3. Hammond, *From Quarks*.
4. See Grosser, *Discovery of Neptune*.
5. Actually, there is a lot more to this fascinating story, including the idea of an inner asteroid belt. See Eisenstaedt, *Curious History*, Chapter 7.
6. Hubble, "Relation."
7. This paper by Riess and others (19 of them) can be found at *http://arxiv.org/abs/astro-ph/9805201*.

## Chapter 2
1. Fuchs, "Amount."
2. Otherwise charge is not conserved; the proton and the electron are equal but oppositely charged, adding to zero, which is the charge of the neutron.
3. Moffat.

## Chapter 3

1.  If, after the first word, you skip five words, and so on, the first paragraph reads: *You should buy* The Unknown Universe *today. The read will be great. (***You*** have been very patient and* ***should*** *be rewarded. To make you* ***buy*** *this, I have a surprise,* ***the*** *kind that has a sneaky* ***unknown*** *clue hidden somewhere in the* ***Universe***, *but you must find it* ***today***. *Sometimes you should read between* ***the*** *lines, and sometimes you must* ***read*** *between the words. Sometimes, you* ***will*** *even totally skip five words.* ***Be*** *observant, and you will see* ***great*** *things happen.)*
2.  Nagano and Watson, "Observations."

## Chapter 4

1.  Barrow, *Infinite*.

## Chapter 6

1.  Hammond, *From Quarks*.
2.  Eisenstaedt, *Curious History*.
3.  For the physicists: There are 16 components of the (nonsymmetric) metric tensor, 24 from the torsion tensor, and 64 from the non-metricity tensor.
4.  I found this translation in Gorelik, "First Steps."

## Chapter 7

1.  Chaisson and McMillan, *Astronomy Today*, 481.
2.  Sublette, "Report."
3.  Klebesadel, "Observations."

## Chapter 8

1.  Anderson, "Positive."
2.  Dirac, "A Theory."
3.  Goudsmit, "Discovery."

## Chapter 9

1.  Timeline of the Big Bang.
2.  Hubble Deep Space Image.
3.  Steinhardt and Turok, *Endless Universe*.

— ✳✳✳ —

# Bibliography

Anderson, Carl. "The Positive Electron." *Physical Review* 43 (1933): 491–494.

Barrow, John. *The Infinite Book*. New York: Random House, 2005.

Chaisson, Eric, and Steve McMillan. *Astronomy Today*. Englewood Cliffs, N.J.: Prentice Hall, 1993.

Dirac, Paul. "A Theory of Electrons and Protons." *Proceedings of the Royal Society of London* 126 (1930): 360–365.

Eisenstaedt, Jean. *The Curious History of Relativity*. Princeton, N.J.: Princeton University Press, 2006.

Fuchs, Burkhard. "The Amount of Dark Matter in Spiral Galaxies." *http://arxiv.org/PS_cache/astro-ph/pdf/0010/0010358v2.pdf* (accessed March 2008).

Gorelik, Gennady. "First Steps of Quantum Gravity and the Planck Values," in *Studies in the History of General Relativity, Einstein Studies, Vol. 3*. Boston: Birkhaeuser, 1992, 364–379. *http://people.bu.edu/gorelik/cGh_FirstSteps92_MPB_36/cGh_FirstSteps92_text.htm* (accessed March 2008).

Goudsmit, S.A. "The Discovery of the Electron Spin." *www.lorentz.leidenuniv.nl/history/spin/goudsmit.html* (accessed March 2008).

Greene, Brian. *The Fabric of the Cosmos*. New York: Vintage Books/Random House, 2004.

Grosser, Morton. *The Discovery of Neptune*. New York: Dover, 1962 (New Edition 1979).

Halpern, Paul, and Paul Wesson. *Brave New Universe*. Washington, D.C.: Joseph Henry Press, 2006.

Hammond, Richard. *From Quarks to Black Holes: Interviewing the Universe*. Singapore: Scientific Press, 2001.

Hawking, Stephen. *A Brief History of Time*. New York: Bantam Books, 1988.

Hubble Deep Space Image, from *http://230nsc1.phy-astr.gsu.edu/hbase/astro/hubdeep.html* (accessed March 2008).

Hubble, Edwin. "A Relation Between Distance and Radical Velocity Among Extra-Galactic Nebulae." From the Proceedings of the National Academy of Sciences Volume 15: March 15, 1929, Number 3. *http://antwrp.gsfc.nasa.gov/diamond_jubilee/1996/hub_1929.html* (accessed March 2008).

Klebesadel, Ray, et al. "Observations of Gamma-Ray Bursts of Cosmic Origin." *Astrophysical Journal* 182 (1973): L85–L88.

Lederman, Leon. *The God Particle*. Boston: Houghton Mifflin Company, 1993.

Lincoln, Don. *Understanding the Universe*. Singapore: Scientific Press, 2004.

Moffat, John. *http://en.wikipedia.org/wiki/John_Moffat_(physicist)* (accessed March 2008).

Nagano, M., and A. Watson. "Observations and implications of the ultrahigh-energy cosmic rays." *Reviews of Modern Physics* 72, (2000): 689–732.

Riess, Adam, et al. "Observational Evidence from Supernovae for an Accelerating Universe and a Cosmological Constant." *http://arxiv.org/abs/astro-ph/9805201* (accessed March 2008).

Steinhardt, Paul, and Neil Turok. *Endless Universe: Beyond the Big Bang*. New York: Doubleday, 2007.

Sublette, Carey. "Report on the 1979 Vela Incident." *http://nuclearweaponarchive.org/Safrica/Vela.html* (accessed March 2008).

Timeline of the Big Bang. *http://en.wikipedia.org/wiki/Cosmological_timeline* (accessed March 2008).

— ✳✳✳ —

# Suggested Reading

When not doing physics I am usually mowing the grass, and do not read very many popular-level books. For this reason there are not many I can recommend, although I know there are many good ones. Here are a few I have read that are great.

Eisenstaedt, Jean. *The Curious History of Relativity*. Princeton, N.J.: Princeton University Press, 2006.

Greene, Brian. *The Fabric of the Cosmos*. New York: Vintage Books/Random House, 2004.

Halpern, Paul, and Paul Wesson. *Brave New Universe*. Washington, D.C.: Joseph Henry Press, 2006. (By the way, Joseph Henry was a great American physicist at a time when they were rare. In his honor, the Henry was adopted as a unit of inductance, a property that characterizes coils of wire.)

Hammond, Richard. *From Quarks to Black Holes*: *Interviewing the Universe*. Singapore: Scientific Press, 2001.

Hawking, Stephen. *A Brief History of Time*. New York: Bantam Books, 1988.

Lederman, Leon. *The God Particle*. Boston: Houghton Miffin Company, 1993.

Lincoln, Don. *Understanding the Universe*. Singapore: Scientific Press, 2004.

# Index

cartesian coordinates, 185
Cat's Eye Nebula, 173
celestial mechanics, 36
cepheid variables, 33
Cerberus, 237-238
CERN, 98
charge center, 91
Cherenkov radiation, 84
classical
    mechanics, 94-95, 98-101
    physics, 87-89
cloud chamber, 192, 236
collapsar model, 182
Collins, Peter, 177-178
coma cluster, 43
combination, 222
commutation relation, 146-147
Compton, Arthur, 67, 68
Compton scattering, 67
confinement, 118
Copernican Revolution, 12-13
Copernicus, 11-12, 14, 32
correspondence principle, 102
cosmic rays, 61-86
cosmological constant, 125
cosmology, 26-27, 230-232
CPT theorem, 237
creation and annihilation
    operators, 100-101, 108
Creator, the, 225-226
Curtis, Heber, 14, 66
cycloid, 231-232
dark
    energy, 31
    matter, 35-60
determinism, 87
dimensional regularization, 248

Dirac, Paul, 127, 150, 195-197, 205
displacement current, 157
Doppler effect, 15, 18
dressed mass, 105
dynamics,
    gravitational, 41-42
    orbital, 41
eclipses, 144
Eddington, Sir Arthur, 144-145
Ehrenfest, 201-203
Einstein, Albert, 9, 22-25, 30, 51,
    57, 67, 72, 87, 104, 112-113, 125,
    138-139, 144-145, 151, 157-159,
    161, 185, 187, 189-190, 194, 218,
    223, 228-230, 238, 245
Einstein-Rosen bridge, 186
Eisenstadt, 144
electric current, 63
electricity, 88
electrodynamics, 67, 150-151
electromagnetism, 98-99,
    138-139, 158
electron degeneracy, 204
electrons, 51-52, 58, 91-92, 132,
    209-210, 244
electroscope, 85
elementary particles, 114-115,
    122, 193
*Endless Universe*, 233
energy, 13, 53, 60, 69-71, 76-77,
    180-181
    dark, 31
    kinetic, 42, 51, 72, 239-240
    potential, 42
error bars, 79-80
Euclid, 151

— ✳✳✳ —

# About the Author

PROFESSOR RICHARD HAMMOND has published numer-
ous scientific articles in a wide range of fields, from general
relativity to quantum mechanics, and has pioneered a new
theory of gravitation that has won international acclaim. He
has won awards from NASA for his research and teaching,
international awards for research on gravity, and was invited
to Cal Tech's Jet Propulsion Lab to study solar system tests
of Einstein's theory. Dr. Hammond brings the fascinating
world of physics to the public. Hammond is an adjunct professor at the University of North Carolina at Chapel Hill,
and works for the Army Research Office as a theoretical
physicist.